Emerging Frontiers in Industrial and Systems Engineering

Continuous Improvement Series

Series Editors:
Elizabeth A. Cudney and Tina Kanti Agustiady

Additive Manufacturing Change Management
Best Practices
David M. Dietrich, Michael Kenworthy, and Elizabeth A. Cudney

Healthcare Affordability
Motivate People, Improve Processes, and Increase Performance
Paul Walter Odomirok, Sr.

Robust Quality
Powerful Integration of Data Science and Process Engineering
Rajesh Jugulum

Statistical Process Control
A Pragmatic Approach
Stephen Mundwiller

Transforming Organizations
One Process at a Time
Kathryn A. LeRoy

Continuous Improvement, Probability, and Statistics
Using Creative Hands-On Techniques
William Hooper

Affordability
Integrating Value, Customer, and Cost for Continuous Improvement
Paul Walter Odomirok, Sr.

Design for Six Sigma
A Practical Approach through Innovation
Elizabeth A. Cudney and Tina Kanti Agustiady

For more information about this series, please visit: https://www.crcpress.com/Continuous-Improvement-Series/book-series/CONIMPSER

Emerging Frontiers in Industrial and Systems Engineering
Success through Collaboration

Edited by
Harriet B. Nembhard, Elizabeth A. Cudney,
and Katherine M. Coperich

CRC Press
Taylor & Francis Group
Boca Raton London New York

CRC Press is an imprint of the
Taylor & Francis Group, an **informa** business

MATLAB® is a trademark of The MathWorks, Inc. and is used with permission. The MathWorks does not warrant the accuracy of the text or exercises in this book. This book's use or discussion of MATLAB® software or related products does not constitute endorsement or sponsorship by The MathWorks of a particular pedagogical approach or particular use of the MATLAB® software.

CRC Press
Taylor & Francis Group
6000 Broken Sound Parkway NW, Suite 300
Boca Raton, FL 33487-2742

First issued in paperback 2020

© 2019 by Taylor & Francis Group, LLC
CRC Press is an imprint of Taylor & Francis Group, an Informa business

No claim to original U.S. Government works

ISBN-13: 978-1-138-59375-6 (hbk)
ISBN-13: 978-0-367-77960-3 (pbk)

Library of Congress Cataloging-in-Publication Data

Names: Nembhard, Harriet Black, editor. | Cudney, Elizabeth A., editor. | Coperich, Katherine, editor.
Title: Emerging frontiers in industrial and systems engineering : success through collaboration / edited by Harriet B. Nembhard, Elizabeth A. Cudney, and Katherine Coperich.
Description: Boca Raton : Taylor & Francis, a CRC title, part of the Taylor & Francis imprint, a member of the Taylor & Francis Group, the academic division of T&F Informa, plc, 2019. | Series: Continuous improvement series | Includes bibliographical references.
Identifiers: LCCN 2019004824 | ISBN 9781138593756 (hardback : acid-free paper) | ISBN 9780429488030 (e-book)
Subjects: LCSH: Industrial engineering. | Systems engineering. | Academic-industrial collaboration.
Classification: LCC T56.24 .E44 2019 | DDC 620.001/171—dc23
LC record available at https://lccn.loc.gov/2019004824

Visit the Taylor & Francis Web site at
http://www.taylorandfrancis.com

and the CRC Press Web site at
http://www.crcpress.com

Contents

Section I   Partnerships, Frameworks, and Leadership

Section II   Engineering Applications and Case Studies

Section III   Postface

Preface

Why This Book?

The field of Industrial and Systems Engineering has evolved as a major engineering field with interdisciplinary strength drawn from effective utilization, process improvement, optimization, design, and management of complex systems. It is a broad discipline that is important to nearly every attempt to solve problems facing the needs of society and the welfare of humanity. In order to carry this forward, successful collaborations are needed between industry, government, and academia. This volume, *Emerging Frontiers in Industrial and Systems Engineering: Success through Collaboration*, provides timely and useful facets, concepts, and methodologies for building those successful collaborations.

Forces to more closely align industry and academia in terms of research in Industrial and Systems Engineering have started to strengthen. For this reason, the Institute of Industrial and Systems Engineers (IISE) moved from having two parallel conferences—one on academic research and the other on practitioners—to one joint conference to explore and promote the integration. After serving together as general co-chairs for 2018 IISE Annual Conference (Orlando) and program co-chairs for 2017 IISE Annual Conference (Pittsburgh), it became clear to us that more could be done to help secure such ties. We conceived and structured this book project through a series of innovation sessions at these conferences.

Many of the authors were participants in those sessions and were instrumental in helping to push the project forward. We hope this book will become a widely used resource throughout our community as well as a template for future volumes to continue to drive success through collaboration.

Harriet B. Nembhard, PhD
Elizabeth A. Cudney, PhD
Katherine M. Coperich

MATLAB® is a registered trademark of The MathWorks, Inc. For product information,

please contact:
The MathWorks, Inc.
3 Apple Hill Drive
Natick, MA 01760-2098 USA
Tel: 508-647-7000
Fax: 508-647-7001
E-mail: info@mathworks.com
Web: www.mathworks.com

Editors

Harriet B. Nembhard is the Eric R. Smith professor of engineering and head of the School of Mechanical, Industrial and Manufacturing Engineering at Oregon State University. Her scholarship is focused on improving complex systems across manufacturing and healthcare. It has led to several advances including a patented imprint lithography manufacturing process for small-scale medical devices and a sensor-based system to conduct early screening for Parkinson's disease. She is an author of a groundbreaking textbook, *Healthcare Systems Engineering*, and of more than 50 journal publications. She has advised 17 PhD candidates and postdoctoral researchers and has been an active mentor to positively impact STEM education, equity and inclusion, and global engagement. Her work has been recognized by election as a Fellow of the Institute of Industrial and Systems Engineers and as a Fellow of the American Society for Quality. She is also a member of the American Society for Engineering Education and the American Society of Mechanical Engineers.

Elizabeth A. Cudney is an associate professor in the Engineering Management and Systems Engineering Department at Missouri University of Science and Technology. She received her BS in industrial engineering from North Carolina State University, her MS in mechanical engineering and master of business administration from the University of Hartford, and her doctorate in engineering management from the University of Missouri–Rolla. In 2018, she was elected as an Institute of Industrial and Systems Engineers (IISE) Fellow, and she received the Crosby Medal from the American Society for Quality (ASQ). In 2014, Dr. Cudney was elected as an American Society for Engineering Management Fellow. In 2013, she was elected as an ASQ Fellow. In 2010, she was inducted into the International Academy for Quality. She received the 2008 ASQ Armand V. Feigenbaum Medal and the 2006 SME Outstanding Young Manufacturing Engineering Award. She has published eight books and more than 80 journal papers. She is an ASQ Certified Quality Engineer, manager of Quality/Operational Excellence, and Certified Six Sigma Black Belt. She is a member of the American Society for Engineering Education, ASQ, IISE, and the Japan Quality Engineering Society (JQES).

Katherine M. Coperich is the manager of external strategic planning at FedEx Ground. She leads the customer facing concept design work for revenue-generating initiatives. She received her BS in industrial engineering from the University of Pittsburgh and is currently pursuing her master of business administration from Temple University. Ms Coperich is a member of the Institute of Industrial and Systems Engineers (IISE) where she participates at both the local and the national level. She is passionate about broadening the reach of science, technology, engineering and math (STEM) within her community and her workplace.

Nembhard, Cudney, and Coperich served together as general co-chairs for 2018 IISE Annual Conference (Orlando) and program co-chairs for 2017 IISE Annual Conference (Pittsburgh).

List of Contributors

Arlethe Yarí Aguilar-Villarreal
Facultad de Ciencias Químicas
Universidad Autónoma de Nuevo
    León
San Nicolás de los Garza, Mexico

Wendell Aldrich
OJI Intertech, Inc.
North Manchester, Indiana

Reza Amindarbari
Center for Geospatial Analytics
College of Design
North Carolina State University
Raleigh, North Carolina

Fiona M. Baxter
Industry Expansion Solutions
North Carolina State University
Raleigh, North Carolina

Mario G. Beruvides
Department of Industrial,
    Manufacturing & Systems
    Engineering
Texas Tech University
Lubbock, Texas

Sam Brannon
Hewlett-Packard Inc.
Corvallis, Oregon

Haleh S. Byrne
Industry Expansion Solutions
North Carolina State University
Raleigh, North Carolina

Sherrie Caltagirone
Global Emancipation Network
Woodinville, Washington

David S. Cochran
Center of Excellence in Systems
    Engineering
Purdue University Fort Wayne
Fort Wayne, Indiana

Paul H. Cohen
Edward P. Fitts Department
    of Industrial and Systems
    Engineering
North Carolina State University
Raleigh, North Carolina

Katherine M. Coperich
Strategic Planning
FedEx Ground
Moon Twp, Pennsylvania

Elizabeth A. Cudney
Department of Engineering
    Management and Systems
    Engineering
Missouri University of Science and
    Technology
Rolla, Missouri

Joel P. Dittmer
Department of Arts, Languages &
    Philosophy
Missouri University of Science and
    Technology
Rolla, Missouri

Todd Easton
Department of Industrial and
    Manufacturing Systems
    Engineering
Kansas State University
Manhattan, Kansas

Jack Feng
Operational Excellence and CVG
  Digital
Commercial Vehicle Group, Inc.
New Albany, Ohio

Walt Garvin
Lean Six Sigma and Industrial
  Engineering
Jabil Inc.
St. Petersburg, Florida

Fernando Gonzalez-Aleu
Department of Engineering
Universidad de Monterrey
San Pedro Garza García, Mexico

Gina Guillaume-Joseph
MITRE Corporation
McLean, Virginia

Aaron Highley
OJI Intertech, Inc.
North Manchester, Indiana

Andrea M. Jackman
IBM Corporation
Armonk, New York

Alireza Jamalipour
The Connecticut Department of
  Transportation
Newington, Connecticut

Dheeraj Kayarat
Micron Technology Inc.
Boise, Idaho

Carl Kirpes
Marathon Petroleum Corporation
Findlay, Ohio

Maria Mayorga
Edward P. Fitts Department of
  Industrial and Systems
  Engineering
North Carolina State University
Raleigh, North Carolina

Patrick McNeff
School of Mechanical, Industrial
  and Manufacturing Engineering
Oregon State University
Corvallis, Oregon

Amin Mirkouei
College of Engineering
University of Idaho
Moscow, Idaho

Daniel Ulises Moreno-Sánchez
Division of Extension, Consulting
  and Research
Universidad de Monterrey
San Pedro Garza García, Mexico

Harriet B. Nembhard
School of Mechanical, Industrial
  and Manufacturing Engineering
Oregon State University
Corvallis, Oregon

Seyed A. Niknam
Department of Industrial
  Engineering and Engineering
  Management
Western New England University
Springfield, Massachusetts

Michael O'Halloran
CH2M
Englewood, Colorado

José Rafael Padilla Valenzuela
Heineken International
Amsterdam, the Netherlands

Brian K. Paul
School of Mechanical, Industrial
and Manufacturing Engineering
Oregon State University
Corvallis, Oregon

Catherine Robertson
Department of Engineering
Universidad de Monterrey
San Pedro Garza García, Mexico

Angela D. Robinson
International Council on Systems
Engineering
San Diego, California

Federica Robinson-Bryant
Embry-Riddle Aeronautical
University-Worldwide
Daytona Beach, Florida

William P. Schonberg
Department of Civil,
Architectural, and
Environmental Engineering
Missouri University of Science and
Technology
Rolla, Missouri

Richard Sereno
OJI Intertech, Inc.
North Manchester, Indiana

Scott Sink
Department of Integrated Systems
Engineering
The Ohio State University
Columbus, Ohio

Kevin C. Skibiski
Missouri University of Science and
Technology
Rolla, Missouri

Dave Sly
Department of Industrial and
Manufacturing Systems
Engineering
Iowa State University
Ames, Iowa

Joseph Smith
Center of Excellence in Systems
Engineering
Purdue University Fort Wayne
Fort Wayne, Indiana

Alice Squires
Department of Engineering and
Technology Management
Washington State University
Pullman, Washington

Laura Tateosian
Center for Geospatial Analytics
North Carolina State University
Raleigh, North Carolina

J. Alexis Torrecillas-Salazar
Department of Engineering
Universidad de Monterrey
San Pedro Garza García, Mexico

Luz M. Valdez-de la Rosa
Department of Engineering
Universidad de Monterrey
San Pedro Garza García, Mexico

Jesus Vazquez
Department of Engineering
Universidad de Monterrey
San Pedro Garza García, Mexico

German Velasquez
Edward P. Fitts Department
of Industrial and Systems
Engineering
North Carolina State University
Raleigh, North Carolina

Teresa Verduzco-Garza
Department of Engineering
Universidad de Monterrey
San Pedro Garza García, Mexico

Shanon Wooden
Department of Industrial
    Engineering & Management
    Systems
University of Central Florida
Orlando, Florida

Section I

Partnerships, Frameworks, and Leadership

1

Remarks on Part I—Partnerships, Frameworks, and Leadership

Harriet B. Nembhard
Oregon State University

Elizabeth A. Cudney
Missouri University of Science and Technology

Katherine M. Coperich
FedEx Ground

Success is driven through collaboration. This book brings together a group of distinguished practitioners and academics to whose work exemplifies successful collaborations. This work spans manufacturing, healthcare, logistics, energy, and several others sectors and applications. This book project arose out of a recognized need to explore the infrastructure, methods, and models for continuing and expanding the space for industrial and systems engineers (ISEs) with respect to academia and practice integration. Section I highlights some of the ways partnerships emerge between those seeking to innovate and educate in industrial and systems engineering, some useful frameworks and methodologies, as well as some of the ideas and practices that undergird leadership in the profession.

Chapter 2 presents a model that expands on capstone design experience to include a full spectrum of opportunities that support universities and industry working together to give students opportunities for experiential learning, leadership, product design, and entrepreneurship. It also points to needs for curriculum development and strategies to expand economic prosperities—needs that may be quite symbiotic in some cases.

Chapter 3 continues to make the case for curriculum development—this time from the standpoint of meeting the needs of Industry 4.0, a term that refers to the fourth industrial revolution and the potential for competitive advantages that stem from technological transformation. It presents a scan of the global landscape and how many countries are making national investments to support high tech including advanced manufacturing, automation, and cyber-physical systems.

The engineering profession is built upon a code of ethics that helps to ensure value to humankind. According to Section I.6 of the National Society of Professional Engineers Code of Ethics for Engineers, "Engineers in the fulfillment of their professional duties shall conduct themselves honorably, responsibly, ethically, and lawfully so as to enhance the honor, reputation, and usefulness of the profession." Chapter 4 provides models for critical developments in ethical design, cross-disciplinary instructional methods for ethics as well as case studies involving ethics violations.

An important dimension of ethical responsibility lies in our capacity to build an inclusive academic community. Doing so will directly affect our students' professional success and their ability to be leaders in the field. Chapter 5 looks at the history and research on women in engineering, and surveys those in ISE discipline on why they chose engineering, why they continue to persist in engineering, and their outlook on having success in engineering. It is perhaps challenging to examine these voices, but doing so will be helpful in improving the system in a holistic way.

Leadership is also driven by a disciplined approach to operational excellence and continuous improvement. Chapter 6 focuses on such methods and especially how they can be designed to support the professional preparation of ISE as undergraduate students and as they commence or extend their careers.

Chapter 7 presents a methodology for supporting sustaining enterprise systems. Motivated by a design of systems structure, the 12-step Collective System Design (CSD) uses lean thinking, lean tools, and lean philosophy to provide an organizational road map that will help meet customer requirements. As with other lean-based approaches, CSD underscores that sustainability comes from embedding it into the organization culture and actions.

Finally, Chapter 8 looks toward emerging technologies for cloud-based services and solutions to streamline and unify the application of ISE methodologies and tools. Efficient ways to deploy ISE will help to drive value creation in organizations and drive collaboration in academia. The cloud-enabled bill of process is used to demonstrate the art of the possible and make the case for higher performance.

2

A Model for Industry/ University Partnerships

Haleh S. Byrne, Paul H. Cohen, and Fiona M. Baxter
North Carolina State University

CONTENTS

2.1 Introduction and Motivation

Universities and industries have been working together to train future workers for more than a century, "but the rise of a global knowledge economy has intensified the need for strategic partnerships that go beyond the traditional funding of discrete research projects. World-class research universities are at the forefront of pioneering such partnerships" (Edmondson, 2012). Today, academic and industry partners are seeking new opportunities to enhance and utilize students' knowledge and skills while providing them with opportunities to lead, serve, take responsibility, and solve complex problems. Rapid technological development and advancement of process efficiency techniques have contributed to the need for strategic partnerships to expand beyond their traditional research roots. In essence, the driver for such strategic collaboration has become more reflective of the need to stay competitive with enhanced diverse utilization of resources available through the university. A robust engineering education curriculum requires students to learn the fundamental subject knowledge that they can apply to real-world projects.

In response to requests from small and medium-sized businesses, both Industry Expansion Solutions (IES) and the Edward P. Fitts Department of Industrial and Systems Engineering (ISE) at North Carolina State University (NCSU) have developed means to work together to synergistically serve the interests of both North Carolina businesses and NCSU engineering students. This chapter outlines the ways in which IES and ISE have collaborated to serve a diverse set of industries in North Carolina, including biotechnology, energy, food processing, manufacturing, and healthcare. This collaboration has fostered both student learning and enhanced industrial productivity by leveraging university resources.

2.2 Missions of NCSU, IES, and ISE

As a Research I, land-grant university, NCSU is committed to serving the people and industries of North Carolina and beyond. As approved by its board of trustees in 2011, the University's mission statement is as follows: "As a research-extensive land-grant university, North Carolina State University is dedicated to excellent teaching, the creation and application of knowledge, and engagement with public and private partners. By uniting our strength in science and technology with a commitment to excellence in a comprehensive range of disciplines, NCSU promotes an integrated approach to problem solving that transforms lives and provides leadership for social, economic, and technological development across North Carolina and around the world."

NCSU demonstrates its commitment to industry engagement, in part, through IES. IES is the engineering-based, solutions-driven, client-focused extension unit of the University's College of Engineering. It serves as a conduit between industry and academia by connecting small and medium-sized businesses with the resources they need to grow and prosper. With a statewide presence, leading experts and industry specialists engineer solutions, sometimes with students, to increase productivity, promote growth through innovation, and enhance performance in support of the NCSU mission. IES is funded by a combination of federal and state support as well as participating industry.

Since 1995, IES has administered the National Institute of Standards and Technology's (NIST) Manufacturing Extension Partnership (MEP) program in North Carolina. The North Carolina Manufacturing Extension Partnership (NCMEP) has demonstrated a $4.9 billion economic impact on North Carolina manufacturing firms since 2000. The strength of the program is demonstrated through the ongoing success of engaging, assessing, and improving the competitiveness of the state's small and medium-sized manufacturers.

The IES mission is to engineer success for North Carolina businesses one solution at a time by understanding its partners, building long-term relationships, crafting meaningful sustainable solutions, and inspiring continuous learning. This aligns with the goals of other units within the College of Engineering, particularly the Edward P. Fitts Department of ISE. This department's commitment to student success and industry engagement makes it a natural collaborator with IES and its engineering outreach to North Carolina companies.

The Edward P. Fitts Department of ISE was established in 1930 as the 16th industrial engineering program in the nation. The department focuses on five main specializations supported by state-of-the-art laboratories: (i) advanced manufacturing, (ii) health systems engineering, (iii) human systems engineering, (iv) production systems and logistics, and (v) systems analysis and optimization. The mission of the department is to support world-class education, research, and outreach in those five areas. Industry interaction and support is a key piece of the Department's mission and one that is accomplished, in part, through its collaboration with IES.

In addition to educating engineering students, the Edward P. Fitts Department of ISE engages in extensive interactions with industry. This includes industrially funded research and collaboration sponsored by Manufacturing USA—the National Network for Manufacturing Innovation. This network is comprised of Advanced Manufacturing Institutes devoted to increasing the nation's manufacturing competitiveness by transferring technologies, such as clean energy and 3D printing, to U.S. manufacturers. Additionally, the department has developed a Corporate Partners Program to facilitate additional collaboration and promote undergraduate education through senior design projects.

The mission of ISE's Corporate Partners Program is to (i) develop and sustain strong connections between the department and the business community; (ii) build strong collaborations between students, faculty, and corporate partners; and (iii) shape the department's future direction. Companies may join the Program at one of four levels. Memberships provide varying degrees of access to students for the purpose of recruiting employees, as well as project sponsorship and inclusion in an advisory board that assists with understanding corporate needs, monitoring student performance in projects, and maintaining a contemporary curriculum that meets the evolving needs of employers.

Both IES and ISE further their corresponding missions and avoid duplication of effort by involving each other in the projects they lead. By working together, the units maximize their individual efforts to serve NCSU, its engineering students, and North Carolina businesses.

2.3 Overview of State and Regional Industries

North Carolina's economy depends on the success of its businesses. The state's gross domestic product (GDP) is over $510 billion, making it the fourth largest state economy in the nation. From 2013 to 2016, the state experienced the sixth fastest economic growth in the nation with a 14.3% increase. The state's growing economy is diverse with companies of all sizes specializing in a wide range of industries, including biotechnology, energy, food processing, manufacturing, and healthcare (Economic Development Partnership of North Carolina, 2018).

IES and ISE each manage centers that focus specifically on serving manufacturers because the manufacturing sector leads the state in GDP contribution at 20%. Currently, there are over 10,000 establishments and over 450,000 employees in the state's manufacturing industry, ranking North Carolina fifth in the nation for that sector. Combined, chemical and food product manufacturing provide the highest contributions to state GDP and employ the largest number of manufacturing workers in the state. Overall, North Carolina manufacturers are responsible for 21% of the state's total economic output and employ 11% of the state's total workforce.

While large manufacturers play a vital role in the state's economic success, small manufacturers are the backbone of the sector. Of the 10,400 manufacturing companies in North Carolina, 80% have fewer than 50 employees and 51% have fewer than 10. To ensure that smaller manufacturers are positioned to grow and succeed, both IES and ISE are intentional about providing them with resources through their centers, programs, and projects.

While manufacturing is an important sector for North Carolina, IES and ISE also work with many service-based clients. Together, they have completed projects in the retail, insurance, technology, education, public service, and healthcare industries. The impacts of projects delivered to manufacturing clients are reported through NIST MEP, whereas the impacts of service-based projects are collected and reported internally. Both reports inform and drive opportunities to enhance program administration, client satisfaction, and student success.

Whether manufacturing or service based, big or small, North Carolina businesses are a primary beneficiary of NCSU Model for Industry Engagement.

IES and ISE serve North Carolina's large and diverse industry segments by delivering a variety of engineering services to small, medium, and large companies, including start-ups, in industries ranging from manufacturing to healthcare. The smaller companies request these services due to limited access to industrial engineers, while larger companies may simply lack the bandwidth to cover all their needs.

2.4 Model for Industry Engagement

As part of the NCSU College of Engineering, both IES and ISE are pride themselves on working diligently to make it simple for business and industry to partner and access University resources and to prepare students for their careers. As professionals who have spent significant time working in and with industry creating a model to enhance student education while assisting companies has been a continuing priority. To accomplish this, they have developed a model for industry engagement that enhances student success through interdepartmental collaboration, industry projects, senior design projects, certifications, and student development.

2.4.1 IES–ISE Collaboration

Collaboration between ISE and IES has been crucial for both organizations to advance their respective missions. Moreover, it has been beneficial to both NCSU engineering students and the industries they serve. Companies of all sizes are able to apply additional engineering talent to problems that are critical to their production of goods and services and the students who help solve those problems gain industry knowledge and hands-on experience. This collaboration between an academic unit and extension service is a primary distinguishing aspect of our model.

IES and ISE work together to deploy customized projects that meet the unique time and capacity constraints of each company they serve. Based on the estimated duration and scope of a project, the two units reach a mutual agreement on which is best suited to take on the project. Some projects that IES takes on may not be suitable for ISE Senior Design due to length, start and completion dates, or scope. However, ISE students may assist IES personnel with some aspects of those projects, as their schedules and skills allow. Assisting with engineering projects, whether led by IES or ISE, is an excellent way for students to gain additional experience and earn money for tuition and other expenses.

The relationship between IES and ISE is truly reciprocal. Just as ISE provides IES with students who are able to assist with projects in the field, IES provides educational resources to ISE students in the classroom. IES engineering experts assist with the ISE's Senior Design course by providing additional instruction in Lean Manufacturing and Six Sigma. By attending student presentations, these experts are able to certify students for a Lean Six Sigma Yellow Belt. When IES and ISE come together to serve students, industry, and each other in this way, the benefits are substantial and measurable.

2.4.1.1 Incorporating Engineering Students into IES Industry Projects

Through its interactions with industry, IES has recognized that clients are consistently interested in engaging engineering students in their work. This is likely due to their desire to prospect for future employees and gain fresh perspectives unhindered by long-term experiences. Interest has come from a diverse set of manufacturers and service-based companies facing engineering challenges in industries such as biotechnology, energy, food processing, manufacturing, and healthcare. In response to particular requests from small and medium-sized businesses, IES has developed, implemented, and evaluated a model focused on employing engineering students on customized projects. The model for most of these projects includes (i) data collection on-site by students, (ii) data analysis on campus, (iii) reporting the results, and (iv) adjustments to final report per client feedback. This approach is efficient and appealing to students because they are able to fulfill their scholastic responsibilities and analyze data from a remote location at their convenience.

Industry projects that utilize industrial engineering principles typically arise as a result of business expansion, needs for better quality and efficiency, demand fluctuations, human interactions in the production process, and continuous improvement programs. The projects are mutually beneficial for the companies who receive engineering services and the students who assist with the delivery of those services. By working directly with industry, students have the privilege of applying engineering concepts they learned in the classroom to hands-on work they are doing in the production and service delivery facilities. Although the IES program focuses on North Carolina, project opportunities for engineering students have also come from several other states, including Georgia, Virginia, and New York.

The organization has learned over time that engineering students are helpful in carrying out projects that have a defined scope and proper supervision. The supervision is performed by IES Lead Engineer for student projects, who also serves as the primary point of contact for the client throughout the project. Through coordination with the University Human Resource System and a commitment of IES supervision, students are employed to work on these projects at competitive wages while gaining knowledge and experience both in the classroom and in the field. They learn to make connections between engineering theory and application, which contribute to their success both as students and future employees. It is important to note that this is some students' first time working in a real-world industry setting. Therefore, they are required to understand customer expectations and commit to the responsibility of delivering results as expected or better. IES and ISE work in conjunction to equip students with the skills they need to succeed as industry workers. In some of cases, students participating in these projects were hired directly by the clients they served based on their contributions and performance during the project period.

In the past 10 years, IES has supervised more than 60 industry projects and has employed nearly 700 students. Depending on the unique needs of each stakeholder, the projects may take as short as 1 month or as long as 1 year to complete. They generally consist of six to eight students from multidisciplinary engineering departments, which encourage students to collaborate with peers beyond their own discipline. Participating in these projects helps students work in diverse teams; develop critical thinking capabilities; and improve their technical, organizational, and communication skills.

Students also learn how to use various tools to carry out their projects. To determine the current state of the "work content," students use video time study software to develop standard times as the foundation and driver to continuous improvement. For projects that focus on facilities layout improvements, students use computer-aided design (CAD) software to better understand physical spaces and manage them more efficiently. Projects that involve food processing and bottling companies give students opportunities to replicate their environments with simulation software in order to recommend the most optimized scenario. In areas of operations, students use various software tools to develop training modules (video, pictorial, etc.), ergonomic assessments, manufacturing cost estimation (labor, overhead, and materials), and product identification and tracking (via radio frequency identification and barcoding).

Students are selected to participate in IES's industry projects through the NCSU-EPack (career services management system), the Engineering Career Fair (https://www.engr.ncsu.edu/careerfair/), and/or direct communication with engineering schools. The criteria for selection are primarily driven by the particular technical skills set required for a customized project. If a student's skills, interests, and availability align with the needs of the project, he or she is considered for a position on the service delivery team.

As a result of working with teams of knowledgeable IES experts and enthusiastic engineering students on industry projects, the client experiences enhanced organizational performance and is able to identify potential process improvement steps to minimize variation, simplify work processes, increase customer satisfaction, decrease operating costs, and establish a consistent level of quality. Feedback from industry clients helps identify the areas of needed focus and research for future projects. Examples include the need for process documentation and analysis and optimization.

2.4.1.2 Assisting Companies through ISE Senior Design Projects

ISE's capstone senior design course seeks to provide companies with professional level engineering solutions to solve critical needs. Over the past 5 years, ISE has managed 88 projects for 60 companies through this course.

Students are attracted to this course because it presents opportunities for creative engineering work in a team setting. Students participate in team-based systems analysis and design, including problem identification and definition, synthesis, specification, and presentation of a designed solution. Essential elements include ISE skills and tools, teamwork, problem solving, and communication. By the end of the course, the students are able to

1. Appropriately apply industrial engineering tools to solve a large, complex real-world engineering problem.
2. Effectively communicate design progress and challenges to project teammates and stakeholders using a variety of methods (email, formal/informal presentation, status charts, and written reports).
3. Develop a design a roadmap that integrates project management and Lean Six Sigma methodologies.
4. Evaluate engineering solutions to determine a recommended design, model, process change, etc. that meets the project's objective(s).

The capstone senior design experience also benefits companies by providing them with a fresh look at existing issues or additional resources that can use to tackle important issues. Projects are solicited from companies and non-profits of all sizes though ISE's Corporate Partners Program. Social media, corporate and alumni databases, and personal contacts lead to most of the projects, but word of mouth and the ISE Department website also inform potential clients about the engineering services they can receive through senior design projects.

ISE encourages senior design projects that represent the broad scope of the ISE discipline. This gives students an opportunity to gain real-world experience using practical tools and performing work in an industry environment that interests them. Common tools that students use to carry out their projects include discrete event simulation, scheduling, automation and robotics, plant layout, ergonomics, optimization, applied statistics, and database systems. Familiarity with these tools further increases their skill sets and employability. In many cases, the students are hired directly by the company they worked with upon graduation. The industry segments that have hired ISE students are shown in Table 2.1.

When selecting and defining a project, companies provide a brief presentation that includes an introduction, background on their products or services, problem definitions and metrics for success, and expected deliverables. These are presented to the class by a company representative or instructor familiar with the company. Students then rank their top three projects and course instructors based on their preferences, schedules, and, to the extent possible, personality and innovation traits.

All projects are completed by groups of three to five students within one semester, although some have required a second part to be completed by

TABLE 2.1

Selected ISE Senior Project Topics by Industry Segment

Industry Segment	Project Topics
Energy	• Scalable, cost-conscious building block model of a sustainable digital infrastructure system
Healthcare	• Development of apps to manage inventories and distribution of over the counter medications • Healthcare analytics • Clinic design and optimization
Information technology and informatics	• Medical record keeping • Development of electronic system for quality • Data mined and identified the leading, lagging, and intangible costs of any particular safety or quality incident; developed a tool that management will be able to use to benchmark performance against industry competitors
Pharmaceuticals, cosmetics, and food processing	• Changeover time minimization • Using value stream mapping techniques, the ISE team will develop a list of changes to improve the system. The team's goal is to reduce process flow time, number of product touches, and travel distance, as well as regain pallet storage racks. This will be beneficial in controlling inventory and reducing costs
Manufacturing	• Flow optimization and waste elimination • Design of robotic welding system • Upholstery processes improvement and a redesign of in-process transportation methods • Part storage, ordering methods, and rules to develop lot size and reorder models that optimize part availability while minimizing carrying costs • Optimized maintenance schedule and improve process productivity by developing a real-time process monitoring system • Automated system to monitor, collect, and report vital process information, and control temperature as well as improve quality performance • Production line flow and balance analyses
Warehousing and logistics	• Standardized processes and established best-practice policies related to these activities in an effort to save on warehousing costs, reduce and eliminate safety risks, and enhance intra-facility communications • Fulfillment center design and control

a new student group in the subsequent semester. Throughout the project period, class meetings are held for (i) presentation of "soft skills" identified in the gap analysis, (ii) group working sessions with faculty mentors, (iii) student presentations, and (iv) guest speakers. Student groups are jointly mentored by an industry representative and faculty instructor.

The course format has several graded portions, each with a primary purpose, as shown in Table 2.2. Reports are written with their industry sponsor in mind, and presentations are considered to be a critical part of the course.

TABLE 2.2

Senior Design Course Requirements

Item	Shared with Sponsor	Primary Purpose
Written proposal and presentation	Yes	Provides company and instructor with clear indication that the group understands and has a plan for project success. Class sees a broad range of applications
In-class assignments	No	Allows students to display mastery of Lean Six Sigma Techniques applied to their project. Allows structured progress toward solution
Project program reviews	Yes	Provides company and instructor with understanding of progress and direction. Feedback provided by instructors, communication specialists, and IES industry specialists
Project reflections and surveys	No	Allows students to step back and reflect on what they have learned and applied
Final report, presentation, and poster	Yes	Serves to provide final solution and recommendations to sponsor. Company presentation allows broad understanding of recommendations. Class sees how other groups have used ISE tools to solve their industry problems. Feedback provided by instructors, communication specialists, and IES industry specialists
Individual contribution	–	Allows instructor, with input from sponsor, to distinguish individual versus group accomplishment

One distinguishing characteristic of ISE's Senior Design course is its engagement of a local communications company to assist students with their project presentations. Specifically, this company provides instruction on creating and delivering effective presentations and joins ISE instructors and IES engineers in providing individual feedback and grading on the effectiveness of each presentation. Feedback from the Senior Design advisory board confirms these strategies to be very effective in improving the quality of students' oral presentations.

To further improve students' presentation and public speaking skills, College of Engineering (COE) departments partner to host a Design Showcase at the end of each semester. ISE invites industry representatives and faculty members to serve as judges for awards that are presented at the Showcase. The hosts also invite the public to attend the students' poster presentations. Creating opportunities for students to present to both general and technical audiences further enhances students' communication skills. By the end of the Showcase, students are better equipped with the skills and confidence they need to deliver effective on-site presentations to sponsors and speak comfortably with potential employers. The incorporation of industry feedback through IES engineers with extensive experience and communications specialists at presentations is a further distinguishing feature of the capstone design experience.

2.4.1.3 Providing Students with Industry Certifications

In 2011, the Edward P. Fitts Department sought to upgrade students' skills based on industry feedback. IES performed a gap analysis on the department's Lean Six Sigma certification, which resulted in some course revisions. For instance, more tools, mostly nontechnical, were added to the Senior Design course. These new tools, coupled with a successful project, enable students to receive a Six Sigma Yellow Belt Certification. An IES representative attends and critiques project presentations to ensure that students are deserving of this professional recognition.

ISE offers a similar certification for students interested in health systems through the Healthcare Engineering Program. Students who select applicable elective courses and complete a senior design project with a healthcare organization are eligible for this certification. This project course is run separately from the manufacturing-based course but with similar objectives.

2.4.1.4 Developing Students to Transition to Industry

While the primary objective for NCSU's engineering projects is to support clients and to deliver results in support of our industry and community, such endeavors have proven to enhance students' technical and soft skills. While technical skills are learned primarily through classroom instruction and hands-on experience, soft skills are instilled through coaching from leaders in both academic and industry settings. This is beneficial for the students and the employers because it empowers students to develop a sense of personal responsibility and independence that they can apply in their careers.

Different students will require different amounts and kinds of attention, advice, information, and encouragement. Since the student teams consist of members with diverse education levels (undergraduate, masters, PhD), there are additional team-building components required to assure the cohesiveness of these units. Some students will feel comfortable approaching their supervisors; others will be shy, intimidated, or reluctant to seek help. A good coach is approachable and available to all students.

The Project Lead, housed at IES, is an experienced engineer who strives to serve as both a supervisor and a coach for the student teams involved. This position can also play a vital role in the students' personal, social, and professional development, as needed. Students are offered the encouraging support and constructive feedback they need as they transition from the University to the workforce.

As best practices, NCSU student project contracts require that the Project Lead be on-site with students throughout the project. This consequently gives students consistent accessibility to their NCSU supervisor. When discussing the characteristics of a good internship for college students, Rangan and Natarajarathinam (2014) state, "[i]t is important to assign a supervisor to guide the intern, help them feel like a part of your team. The supervisor should also have frequent meetings with the intern to provide feedback and direction." The Project Lead applies these principles when serving students in this multifaceted role.

NCSU is diligent about appointing supervisors to its engineering students as a way to support and instill them with invaluable soft skills, but there are opportunities for companies to aid universities in shaping students into well-rounded, industry-ready engineers. "It is important that the business assigns mentors to provide direction, feedback and recognition through the course of the program" (Rangan and Natarajarathinam 2014). The role of business in engaging and guiding the students during projects cannot be understated. Although a typical IES project is more ad hoc in its on-site business activity, it is certainly advantageous for the client to get involved in nurturing the student workers, especially when future recruiting opportunities are considered.

2.5 Results of Industry Engagement

The NCSU Model for Industry Engagement provides a gateway for the University and North Carolina industry to work together to ensure the state is equipped with a highly skilled and knowledgeable workforce. By seeking to understand industry needs and incorporating feedback into its engineering curricula and projects, NCSU can assure that its engineering graduates are well prepared to make valuable contributions to small, medium, and large companies within North Carolina's large and diverse industry sectors. In order to demonstrate the success of the project and continuously improve our efforts, a formative and summative evaluation model measures both industry and student outcomes.

2.5.1 Industry Outcomes

While IES communicates directly with clients during and upon completion of the industry projects, there are additional ways to quantify long-term impacts of project outcomes. Every quarter, IES participates in a manufacturing client survey process for completed projects. These surveys are administered through the NIST MEP program and seek to collect important information from clients on the impact of services provided using an independent third-party vendor. This survey's purpose is to use the client-reported outcomes in areas such as new and retained sales, cost savings, investments, and new and retained jobs to estimate the overall effect of MEPs on the U.S. economy. In fiscal year 2017, NCMEP clients reported $721,586,470 in economic impact. Student projects ($887,000 project cost) resulted in $79,859,323 in reported economic impact and 439 created or retained jobs.

TABLE 2.3

Industry Comments Regarding Quality of Work, Student Preparation, and Performance

"Great service and the student was outstanding. Great to get help that is talented and self-motivated...."

"Great services and very professional. We would highly recommend to others."

"...very responsive and easy to work with and efficient. We hired a student full time for the summer and then part time during the fall....[and]...full time when he graduated..."

"We found the student on demand program very effective for our project."

"Great service and the student was outstanding. Great to get help that is talented and self-motivated to complete the tasks given in a timely manner."

"...team did a great job understanding, gathering and presenting options that we could implement..."

"...students have shown themselves to be intelligent, adaptable, and have strong aptitude with engineering principals."

"Great attitudes and team interactions..."

"Very technically capable and knowledgeable, particularly with specific tools used in the execution of the assignments."

"I have helped judge senior design projects for the ISE Department at NC State for several years and have noted two things. There has been a steady improvement in the quality of the solutions that the students develop, and the projects themselves have evolved to be much better representations of the challenges that new engineers will face in their first career positions."

At multiple points throughout the senior design projects, companies provide feedback on the performance of individual students and the overall project progress and results. Most projects are with companies that have previously sponsored similar projects, indicating the positive impact that student projects have had on their organizations. Project sponsors also serve on the senior design advisory board, where they evaluate the preparation of our students to work with them and provide recommendations for improvement (Table 2.3).

2.5.2 Student Outcomes

Through alumni and company surveys, the learning outcomes have been achieved through ISE Senior Design experiences. Specifically, students are able to (i) apply industrial engineering tools to solve complex real-world problems, (ii) communicate more comfortably and effectively, (iii) integrate project management and Lean Six Sigma tools for successful project completion, and (iv) evaluate multiple engineering solutions for defensible recommendations. Moreover, it is clear from students' written and oral comments that the experience boosts their confidence and helps them perform well in professional interviews upon graduation. From surveys completed by NCSU College of Engineering graduates, it is clear that the Senior Design experience effectively prepares them to transition to an industry environment, as shown in Table 2.4.

TABLE 2.4

Selected Graduating Senior Outcomes Reflecting Critical Abilities

Outcome	Percentage Rating Ability as Excellent or Very Good
Ability to apply knowledge from major	96
Ability to apply mathematics	96
Ability to design systems to meet needs	92
Abilities to function in multidisciplinary teams	92
Ability to identify, solve engineering problems	96

TABLE 2.5

Representative Student Comments Regarding Their Experience in Industry Projects Including the Role of IES

Student Statements
"NCSU IES allowed me to see what "real world problem solving" was like outside the structure of a classroom and curriculum."
"My experience at IES helped me grow personally and professionally."
"NCSU IES consulting work gave me a very good opportunity to apply my knowledge and skill into reality."
"Personally, it was a great experience helping me improve my time management, leadership and communication skills."
"…experience helped me significantly to get my next job as the recruiters were obviously impressed when they learned about the projects I worked on before."
"Not only was I able to utilize my skills but also gain new skill sets."
"…working with IES was an excellent experience for me as it shaped my academic and career life."

Likewise, the IES's industry project experience also has a highly positive influence on students who participate at varying points in their academic careers. Students appreciate the informal learning experiences they receive through real-world work experience and mentoring. As shown in Table 2.5, students' comments on the impact of their experiences reflected the positive influence of applying classroom knowledge to industry practice, acquiring new technical and soft skills and growing on both professional and personal levels. The need and impact of effective mentoring was also noted.

2.6 Conclusions

The model outlined is clearly in line with achieving the goals of both ISE and IES, the College of Engineering, and the NCSU in line with our land-grant mission. The evaluation plan quantifies the program's contribution to the economic development of the state of North Carolina. These measures are further complimented by the firsthand information gathered from student participants who report a variety of ways that their participation in the program has advanced their career and professional goals.

2.6.1 Opportunities for Continuous Improvement

While both IES and ISE programs have been successful, both independently and in collaboration, there are clear opportunities for continuous improvement. For instance, the model that has been developed to navigate a large university and understand the best paths for addressing companies' unique needs can and should be replicated between other units on campus. In addition, engineering students would benefit from earlier exposure to industry experiences in their undergraduate tenure to achieve accelerated professional growth. Earlier exposures coupled with practical internship experiences would better prepare students for employment and produce even better Senior Design results. Recognizing and addressing these opportunities for improvement can benefit both engineering units, their students, and the industries they serve.

From students' comments regarding their participation in engineering projects, it is clear that they desire increased levels of guidance and involvement in a greater number of projects (Table 2.5). Additionally, students cite increasing confidence and personal growth as a key outcome of the coaching and supervision they receive. Therefore, to better serve the NCSU's College of Engineering, its students, and North Carolina industry, the two units will incorporate more ISE faculty in IES engineering projects and more IES personnel in ISE senior design projects moving forward. Realizing and acting on additional opportunities for IES–ISE collaboration will broaden and deepen the impacts the units have synergistically achieved thus far.

2.6.2 Potential for Scale-Up

The model used at NCSU employs the resources of an academic department (ISE) and extension service (IES). A similar collaboration for mutual benefit can be created through collaboration of academic and nonacademic partners, even if not in the same organization. This can include the regional MEP or other organizations with a similar mission. The inclusion of people with communications expertise and industry experience can be included in many ways to enrich the student capstone design experience while providing a service to local industry. Student and industry feedback clearly indicates that this combination prepares students for success.

References

Economic Development Partnership of North Carolina. 2018. "Industries." https://edpnc.com/industries/.

Edmondson, Gail. 2012. Making Industry Partnerships Work: Lessons from Successful Collaborations.

North Carolina State University. The Pathway to the Future: NC State's 2011–2020 Strategic Plan. https://strategicplan.ncsu.edu/pathway-to-the-future/.

Rangan, Sudarsan and Natarajarathinam, Malini. 2014. How to Structure an Internship that Is Great for the Intern and the Manager?

3

Industry 4.0: Success through Collaboration

Fernando Gonzalez-Aleu,
Catherine Robertson, Jesus Vazquez, Teresa Verduzco-Garza,
J. Alexis Torrecillas-Salazar, and Luz M. Valdez-de la Rosa
Universidad de Monterrey

CONTENTS

3.1 Introduction

3.1.1 Industry 4.0

The term "Industry 4.0" became publicly known in 2011, when representatives from business, politics, and academia supported the idea to strengthen the competitiveness of the German manufacturing industry (Kagermann, Lukas, and Wahlster, 2011). Promoters of this idea expect Industry 4.0 to be a change agent to improve different industrial processes, such as manufacturing, engineering, material usage, supply chain, and life cycle management (Kagermann, Helbig, Hellinger, and Wahlster, 2013). Since its inception, Industry 4.0 has gained tremendous interest from governments, public organizations, and private industries. The German government invested USD 200 million to spur Industry 4.0 research across government, academia, and business as part of their "high-tech strategy 2020" initiative

(Zaske, 2015); the UK government committed to codevelop advance manufacturing technology standards (Addison, 2014); the U.S. government initiated the "Smart Manufacturing Leadership Coalition" with an investment of USD 140 million to develop new technology solutions in advanced manufacturing (Anonymous, 2016); and the Chinese government initiated a 10-year government program focusing on automation and cyber-physical systems (CPSs) (Pardo, 2016).

Although Industry 4.0 is currently a top priority for many companies, research centers, universities, and countries, there is variation in the understanding of the term and what it encompasses. Industry 4.0 alludes to a fourth industrial revolution enabled by the following concepts: smart factories, smart services, CPSs, Internet of things, self-organization, new systems in distribution and procurement, new systems in the development of products and services, adaptation to human needs, and corporate social responsibility (Kagermann, Helbig, Hellinger, and Wahlster, 2013; Wollschlaeger, Sauter, and Jasperneite, 2017).

3.1.2 Challenges and Benefits from Industry 4.0

Although the foundation of Industry 4.0 is automation and data exchange, there is evidence suggesting that several organizations are still collecting and analyzing data using paper and pencil (Lyle, 2017). Therefore, in order to increase the number of organizations implementing Industry 4.0, six challenges should be addressed (Grangel-González et al., 2016; Schlick, Stephan, Loskyll, and Lappe, 2014):

- *Interoperability*. Ability to connect CPS, humans, and Industry 4.0 factories via the Internet of things and Internet of Services
- *Virtualization*. Ability to monitor physical processes, linking sensor data (monitoring physical processes) with virtual plant models and simulation models
- *Decentralization*. The ability of CPSs within Industry 4.0 factories to make decisions on their own
- *Real-time capability*. Capability to collect and analyze data and provide the derived insights immediately
- *Service orientation*. Availability of CPSs, the services of companies, and humans to other participants, both internally and across company borders
- *Modularity*. Adaptability of Industry 4.0 factories to changing requirements by replacing or expanding individual modules

Organizations that address these challenges will be able to obtain one or more of the following five benefits: time, cost, flexibility, integration, and sustainability (Schuster et al., 2015; Gabriel and Pessl, 2016). Engineers spend 31% of working time searching for information, time that can be used for activities that produce value. Incorrect information and resulting erroneous decisions cost as much as 25% of the company's income. Only 36% of companies are ready to optimize processes based on data analysis. Companies may reduce up to 80% of the time dealing with production interruptions if they use digital validation to involve the simultaneous development of the product and the production process. Providing detailed information on each point of the production process, resource, and energy use can be optimized over the entire value network.

3.1.3 Impact on Workers

Industry 4.0 experts insist that human work will play a very important role in the future production. There still will be classic manually running processes that have to be accomplished intelligent, creative, or flexible workers (Gabriel and Pessl, 2016). However, the positioning of people within the production and the tasks they have to be performed will change (Ganschar, Gerlach, Hämmerle, Krause, and Schlund, 2013; Schließmann, 2017). The strong network of people and machines within Industry 4.0 mean that job content, work processes, work environment, and the required skills need to be changed. As physical tasks will be less relevant in future, expertise in new communication technologies in terms of planning, execution, and decision-making processes (data collection and analysis); control or programming; and error correction will become increasingly important (Kagermann, Helbig, Hellinger, and Wahlster, 2013). In addition, demand for interdisciplinary knowledge and skills to understand working and thinking will become more essential (Kurz and Metall, 2012). Engineering students should be prepared on the technological concept of CPSs and the Internet of things to collaborate with employees from other disciplines (Schuster et al., 2015). Hence, it is important to understand the current level of collaboration observed in this field.

3.1.4 Chapter Purpose and Research Framework

The purposes of this chapter are (i) to determine the level of collaboration currently observed between authors publishing about Industry 4.0 and (ii) to determine the impact of Industry 4.0 in undergraduate engineering curriculum redesign. To achieve both goals, the research involved three phases. First, a systematic literature review was used to collect relevant publications in this field and assess author characteristics dimension using the following criteria (Keathley-Herring et al., 2016): author quantity and author collaboration. Second, the 2020 Industrial and Systems Engineering and Engineering Management curriculums' redesign process was documented identifying new competencies, new courses, and synergy with other academic departments need to satisfy organizations requirements (including Industry 4.0). Finally, findings, research limitations, and propose future work were examined.

3.2 Industry 4.0: A Systematic Literature Review

3.2.1 Systematic Literature Review Methodology

A systematic literature review is a protocol designed to locate, appraise, and synthesize the best available literature relating to a specific research topic (Tranfield, Denyer, and Smart 2003). A six-step systematic literature review methodology was used to collect relevant literature related to Industry 4.0 (Keathley-Herring et al., 2016):

a. *Problem definition.* The world stands on the threshold of a new age of technology, which will launch a fourth industrial revolution. According to this idea, a web-based network will support smart factories at every stage of the work on the product, from design through to servicing and recycling. It is a vision of a world in which the real environment connects to the digital one using the following driving forces: Internet of things, cloud computing, big data, CPSs, and others. In response to this new demand, engineering students should be prepared on the technological concepts to collaborate with employees from other disciplines (Schuster et al., 2015). Hence, it is important to understand the current level of collaboration observed in this field. On the other hand, this new topic has been attracting attention worldwide, but due to the newness of the topic and the complexity of the Industry 4.0 phenomenon, the available literature lacks efforts to systematically review the state of the art of this new industrial revolution wave (Liao, Deschamps, Loures, and Ramos, 2017). Therefore, the first goal is to determine the level of collaboration currently observed between authors publishing about Industry 4.0 using the Keathley-Herring et al. (2016) systematic literature review framework.

b. *Scoping study.* The research team decided to conduct a precise search instead of a sensitive search in order to identify publications highly related to Industry 4.0. The goal of a sensitive search is to collect most of the publications available mentioning the search terms in all search fields (e.g., full text, instead of title or abstract). In this type of search, it is expected to find several publications not related at all to the main topic of the investigation (Lefebvre, Manheimer, and Glanville, 2011).

c. *Search strategy.* This step of the systematic literature review is integrated into three concepts: databases, search protocol, and delimiters. First, three databases/platforms were used to collect the publications available: Scopus, ProQuest, and EBSCO host. Second, the search protocol is integrated with search terms, Boolean operators, and search field. The search protocol was tested several times and modified until the research team obtained the final version used in the three database/platforms: "Industry 4.0" AND "cyberphysical systems" AND "internet of things" AND ("smart factory" OR "smart manufacturing") in "All fields except full text." Third, publication date and language were the two delimiters applied in this search strategy. The search protocol was conducted during summer 2018; however, the research team decided to limit the search until December 2017 to include in this investigation only information from a complete year. All of the research team members are fluent in English and Spanish and one is fluent in German. This capability gave us the opportunity to collect papers in these three languages.

d. *Exclusion criteria.* Four exclusion criteria were used to remove publications from this investigation: duplicate publications, publications with low content or knowledge (e.g., trade press and working papers), publications not related to Industry 4.0, and publications without full text available.

e. *Data collection.* The following data was collected from the papers that passed the exclusion criteria: publication year, author's name, author's country of affiliation, author's institution and department of affiliation department of affiliation, and author's frequency of publication.

f. *Data analysis and synthesis.* Using the data collected from the publications that passed the exclusion criteria, the following analysis and syntheses were conducted: final publication set, author quantity, and author collaboration. First, final publication set describes integration (e.g., proportion of publications per platforms/databases and frequency of publications per year). Second, author quantity includes the number of existent authors and the analysis of the frequency of new authors per year. Finally, author collaboration includes the analysis of multiple author papers and publications with authors from a different country.

3.2.2 Final Publication Set

A total of 441 publications were identified from the search strategy, but only 281 publications (conference proceedings, journal articles, and book chapters) passed the exclusion criteria (an inclusion rate of 64%). These publications were found in the following databases/platforms: 206 publications in

Scopus (73%), 29 publications in EBSCO host (10%), 22 publications in ProQuest (8%), eight publications in Scopus and ProQuest (3%), eight publications in EBSCO host and ProQuest (3%), five publications in EBSCO host and Scopus (2%), and three publications in the three databases/platforms (1%). This finding highlights the importance of using multiple database/platforms during a systematic literature review.

Since 2013, the first publication found (1 out of 281 publications; 0.4%), the number of publications per year increased (see Table 3.1), suggesting that Industry 4.0 is capturing the attention from practitioners and academics.

It is impressive to note that Industry 4.0 interest reached 43 author countries of affiliation, suggesting a worldwide recognition of the importance of this topic. The top ten author's country of affiliation were China (18.4%), Germany (16.0%), the United States (7.9%), Spain (7.8%), Italy (7.0%), the United Kingdom (6.5%), Korea (3.7%), Brazil (2.7%), Portugal (2.6%), and Taiwan (2.5%).

3.2.3 Author Characteristics Dimension

As was mentioned in Section 3.1, Industry 4.0 requires collaboration between professionals or academics from engineering in different fields. Hence, it is important to analyze author characteristics. According to Keathely-Herring et al. (2016), author characteristics could be assessed using three different criteria: author quantity, author diversity, and author collaboration. In order to achieve the first goal of this chapter, to determine the level of collaboration currently observed between authors publishing about Industry 4.0, the research team selected author quantity and author collaboration criteria.

Author quantity was assessed using two sub-criteria: average authors per publication and frequency of authors per publications. From the 281 publications, 815 unique authors were identified, showing an average of 2.9 authors per publication; indicating collaboration between authors.

TABLE 3.1

Frequency of Publications per Year

Year	Frequency of Publications	%
2013	1	0.4
2014	8	2.8
2015	29	10.3
2016	87	31.0
2017	156	55.5
Total	281	100

A considerable increase in the frequency of publications per year related to Industry 4.0; evidence of the relevance of this topic for academic and/or practitioners are observed.

This finding is reinforced by the frequency of authors per publication (see Table 3.2), where only 27 out of the 281 publications (9.6%) were published by a single author.

Author collaboration was assessed using two sub-criteria: social network analysis by author name and social network analysis by author country of affiliation. In both cases, Gephi 0.9.2 software was used to visualize the social network.

As the research team mentioned previously, the evidence suggests that an emerging group of authors is working together to generate knowledge about Industry 4.0. This situation could be better understood in Figure 3.1; where each node represents a unique author and edges represent the connections between authors from the same publications. As it is observed in Table 3.2 and Figure 3.1, most of the publications were done by teams of authors with two or more members. Also, there are some authors working with two or more different teams (see Figure 3.1), such as Di Li, Gunther Schuh, Hehua Yan, JianFu Wan, and Keliang Zhou, indicating that authors publishing on Industry 4.0 are collaborating.

Collaboration with authors' country of affiliation represents if authors from different countries are working together in Industry 4.0. In this situation, nodes represent each author's country of affiliation and edges represent the

TABLE 3.2

Frequency of Authors per Publication

Number of Authors per Publication	Frequency of Publications	%
1	27	9.6
2	55	19.6
3	63	22.4
4	58	20.6
5	40	14.2
6	25	8.9
7	6	2.1
8	2	0.7
9	2	0.7
10	2	0.7
11	0	0.0
12	0	0.0
13	0	0.0
14	0	0.0
15	1	0.4
Total	281	100

Most of the publications (90.4%) were written by two or more authors, showing a high level of collaboration on this research field.

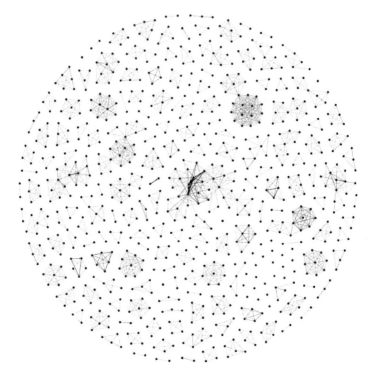

FIGURE 3.1
Authorship social network. In order to increase the knowledge about Industry 4.0 and the quality of this knowledge, it is important to observe collaboration between authors; this is observed in mainly in the center of this figure.

connections between authors from the same publication. Figure 3.2 shows China, Germany, and the United States are authors' country of affiliation with more collaborations: these authors' country of affiliation collaborated with 26 out of 43 different countries detected in this research.

Author quantity and author collaboration criteria show collaboration between authors publishing on Industry 4.0, including the identification of a set of predominant authors and predominant authors' country of affiliation. Now, it is also important to analyze the impact of Industry 4.0 in undergraduate engineering curriculum redesign.

3.3 Redesigning Industrial and Systems Engineering and Engineering Management Curriculums

The Universidad de Monterrey is a Mexican higher education institution with 14,000 students that offers 46 undergraduate degree programs,

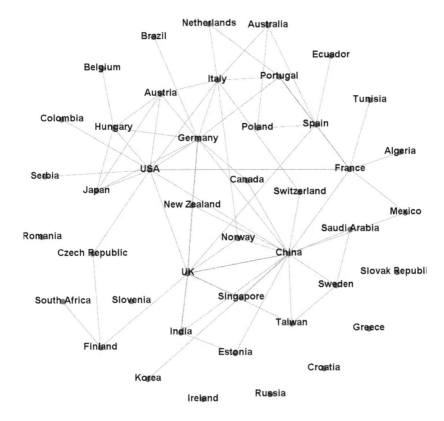

FIGURE 3.2
Social network of authors' country of affiliation. A different way to analyze the social network of authors is observing the collaboration of authors from different countries of affiliation. In this case, the figure shows that authors from China, Germany, the United States, and the United Kingdom are working together and collaborating with authors with other country of affiliation.

15 master degree programs, 15 graduate specialties, and 35 medical specialties across six colleges: art, architecture and design, education and humanities, business, health sciences, law and social sciences, and engineering and technology.

As a response of market demand and government requirements, the 46 undergraduate degree programs update their curriculum that will be offered to freshman students in 2020. The following case study describes how professor, program directors, alumni, government, and organizations work to integrate Industry 4.0 in the new curriculum for Industrial and Systems Engineering and Engineering Management programs.

Each undergraduate program director works using the same methodology that consists in five phases: board of directors, curriculum design requirements, national and international benchmarking, strengths and weakness of the current curriculum, and 2020 curriculum design. First, undergraduate program director integrated five boards (see Table 3.3). Each of these boards had different purposes.

Second, undergraduate program directors collected information from Universidad de Monterrey, government, and national and international accreditations related to a number of credits required in different knowledge areas, as well as student outcomes. As results of this phase, the maximum number of credits authorized for Industrial and Systems Engineering and Engineering Management were 315 and 312 credits, respectively, but Universidad de Monterrey requires that 60 credits should be used in common courses (independently of the bachelor degree awarded, all students from Universidad de Monterrey will take the same 60 credits). Therefore, there are only 255 or 252 credits available to satisfy organizations' needs government and accreditations.

Third, undergraduate programs directors asked faculties board members about the top national and international universities in their field. The curriculums from 20 different universities around the world (the United States,

TABLE 3.3

Industrial and Systems Engineering and Engineering Management Board of Directors Constitution

	Role	Industrial and Systems Engineering	Engineering Management
Faculty board	Collect tendencies in different knowledge areas according to faculty expertise	Nine faculties	Nine faculties
Alumni and board	Collect information about strengths and weakness of the current curriculum	Fifteen alumni	Ten alumni
External board	Collect information about current and future skills, knowledge, and aptitude that students need to develop	Five managers	One independent advisor One cluster director Two entrepreneurs Two operation directors
Sinergy board	Define how expertise from other knowledge areas (business, finance, human resources, etc.) could improve the future curriculum	Four undergraduate and graduate directors at the Universidad de Monterrey	Six undergraduate and graduate directors at the Universidad de Monterrey

The diversity of the board member constitution (faculties, alumni, managers/directors, entrepreneurs, and other Universidad de Monterrey employees) was created with the goal to obtain a complete overview of Industry 4.0 impacts in the current and future professionals.

the United Kingdom, Mexico, Singapore, Australia, and China) were ana-lyzed, identifying main knowledge area offers. Fourth, focus group and personal interviews with alumni board members and external board mem-bers were used to collect strengths and weakness of the current curricu-lum. Information collected from Industrial and Systems Engineering and Engineering Management board of directors were classified by affinity iden-tifying 12 strengths and 24 weakness, for example, system thinking gave us the opportunity to understand how a decision could impact different areas in the organization (strengths); working with employees from differ-ent culture and education levels (strengths); senior capstone projects gave us skills need in our job (strengths); faculty members force us to think in solutions instead of apply formulas (strengths); some courses have similar content (weakness); there are not content related to Industry 4.0 (weakness); there are not courses related to sales, purchasing, and other management skills (weakness); and include the utilization of different software such as Excel and SQL (weakness). Finally, faculty board members and undergradu-ate program directors worked together with the information collected from previous phases to define undergraduate curriculum redesign (see Table 3.4), in reviewing course syllabi, eliminating topics, creating new courses, and upgrading old courses.

In summary, the 2020 Industrial and Systems Engineering curriculum fin-ished with 39% of new credits: 30 out of the 99 new credits were focused to address Industry 4.0 demand. These 30 new credits were also added in the 2020 Engineering Management curriculum and include the incorpora-tion of courses such as acquisition and data structure, data processing, and data/business analytics. During these courses, students should be able to

TABLE 3.4

Industrial and Systems Engineering and Engineering Management Undergraduate Curriculum 2020

Area	Industrial and Systems Engineering		Engineering Management	
	Old Courses (Credits)	New Courses (Credits)	Old Courses (Credits)	New Courses (Credits)
Computer science	0	6	0	6
Physics and mathematics	36	24	18	30
Engineering	120	69	102	72
Management	0	0	18	6
Subtotal	156	99	138	114
Total	255		252	

As a consequence of the impact that Industry 4.0 will have in the professionals, courses from different areas of the curriculum in both undergraduate programs were updated or changed by new courses.

learn software or programing language such as MATLAB, Python, and SQL. Additionally, 12 new credits related to Industry 4.0 were included in the 2020 Engineering Management curriculum (forecast and multivariate analysis); therefore, 40 out of 108 new credits were used to address Industry 4.0 demand.

On the other hand, the integration of computer science, physics and mathematics, and management courses (only for Engineering Management curriculum) shows how Industry 4.0 requires that engineers acquire skill, knowledge, and aptitude outside the engineering field.

3.4 Discussion

According to the experts in Industry 4.0, engineers should reinvent themselves and learn about CPSs and Internet of things, but they also need to be able to collaborate with engineers from different backgrounds in order to succeed in the implementation of Industry 4.0 concepts in more organizations, addressing six challenges: interoperability, virtualization, decentrability, real-time capability, service orientation, and modularity. Therefore, the research team defined two purposes in this chapter: (i) to determine the level of collaboration currently observed between authors publishing on Industry 4.0 and (ii) to determine the impact of Industry 4.0 in undergraduate engineering curriculum redesign. After conducting a systematic literature review and assessed the author characteristics dimension. The frequency of authors per publication, social network analysis by author name, and social network analysis by authors' country of affiliation show the emerging of clusters (collaboration) between experts publishing on Industry 4.0.

Methodologically, this chapter addressed two limitations in order to improve the quality of the findings. First, a systematic literature review does not warranty to collect all the information available on a specific topic. For that reason, the research team used multiple databases/platforms and different search terms to collect several publications. According to the information showed in the result section, 73% of the publications were found in Scopus, 10% in EBSCO host, and 8% in ProQuest. This finding sustains the idea of using several databases/platforms during systematic literature review. Second, the research leader was aware of the relevance of having a research team fluent in two or more languages. The integration of a practitioner fluent in German in this research team was a key factor for the results obtained in this chapter.

On the other hand, Industrial and Systems Engineering and Engineering Management curriculum redesign show that Industry 4.0 had a high impact in the incorporation of new courses, some of these courses outside the engineering field, to develop new skills, knowledge and aptitude, such as structure thinking, data collection, data analysis, and software use. With these courses, Industrial and Systems Engineering and Engineering Management at Universidad de Monterrey will be able to address three out of the six challenges that organization will have in their process to apply Industry 4.0 concepts (virtualization, real-time capability, and service orientation).

However, it is important to highlight that findings from this case study should not be generalized; instead, each university needs to upgrade their engineering curriculum according to market conditions and organizations/ customer demands.

Although this chapter offers relevant information to practitioners and academics with a small knowledge of Industry 4.0, future work should follow these research lines: analyze the collaboration between authors from different departments in an organization (e.g., quality, information technology, marketing, production, maintenance) and conduct empirical investigations to collect information on organizations implementing Industry 4.0 to identify critical success factors.

References

Addison, P. 2014. UK's first digital factory demonstrator launched. www.siemens. co.uk/en/news_press/index/news_archive/2014/uks-first-digital-factory-demonstrator-launched.htm (accessed January 29, 2018).

Anonymous. 2016. Smart manufacturing leadership coalition will lead the new smart manufacturing innovation institute. www.prnewswire.com/news-releases/smart-manufacturing-leadership-coalition-will-lead-the-new-smart-manufacturing-innovation-institute-300287952.html (accessed January 29, 2018).

Gabriel, M. and Pessl, E. 2016. Industry 4.0 and sustainability impacts: Critical discussion of sustainability aspects with a special focus on future of work and ecological consequences. *Annals of the Faculty of Engineering Hunedoara* 14:131.

Ganschar, O., Gerlach, S., Hämmerle, M., Krause, T., and Schlund, S. 2013. *Produktionsarbeit der Zukunft-Industrie 4.0*. Stuttgart: Fraunhofer Verlag.

Grangel-González, I., Halilaj, L., Auer, S., Lohmann, S., Lange, C., and Collarana, D. 2016. An RDF-based approach for implementing Industry 4.0 components with administration shells. In *Emerging Technologies and Factory Automation (ETFA), 2016 IEEE 21st International Conference on*, Berlin, Germany. IEEE, September 6–9, pp. 1–8.

Kagermann, H., Lukas, W. D., and Wahlster, W. 2011. Industrie 4.0: Mit dem Internet der Dinge auf dem Weg zur 4. industriellen Revolution. *VDI nachrichten* 13:11.

Kagermann, H., Helbig, J., Hellinger, A., and Wahlster, W. 2013. Recommendations for implementing the strategic initiative Industrie 4.0: Securing the future of German manufacturing industry; final report of the Industrie 4.0 Working Group. Forschungsunion.

Keathley-Herring, H., Van Aken, E. M., Gonzalez-Aleu, F., Deschamps, F., Letens, G., Orlandini, P. C. 2016. Assessing the maturity of a research area: Bibliometric review and proposed framework. *Scientometrics* 109:927–951.

Kurz, C. and Metall, I. 2012. Arbeit in der Industrie 4.0. *Information Management and Consulting* 3:56–59.

Lefebvre, C., Manheimer, E., and Glanville, J. 2011. Searching for studies. In *Cochrane Handbook for Systematic Reviews of Interventions*, Higgins, J. P. T. and Green, S. eds., pp. 95–150. West Sussex: John Wiley & Sons Ltd.

Liao, Y., Deschamps, F., Loures, E. F. R., and Ramos, L. F. P. 2017. Past, present and future of Industry 4.0-a systematic literature review and research agenda proposal. *International Journal of Production Research* 55:3609–3629.

Lyle, M. 2017. From paper and pencil to Industry 4.0: Revealing the value of data through quality intelligence: For manufacturers to realize the full power of quality and the data behind it, they must embrace the technology available. *Quality* 56:25–30.

Pardo, N. 2016. China, Asia carve out stake in Industry 4.0. www.ptc.com/en/product-lifecycle-report/china-asia-carve-out-stake-in-industry-4-0 (accessed January 29, 2018).

Schlick, J., Stephan, P., Loskyll, M., and Lappe, D. 2014. Industrie 4.0 in der praktischen Anwendung. In *Industrie 4.0 in Produktion, Automatisierung und Logistik*, pp. 57–84. Wiesbaden: Springer Vieweg.

Schließmann, A. 2017. iProduction, die Mensch-Maschine-Kommunikation in der Smart Factory. In *Handbuch Industrie 4.0*, Vogel-Heuser, B., Bauernhansl, T., and ten Hompel, T. eds., Bd. 4, pp. 173–202. Berlin, Heidelberg: Springer Vieweg.

Schuster, K., Plumanns, L., Groß, K., Vossen, R., Richert, A., and Jeschke, S. 2015. Preparing for Industry 4.0—Testing collaborative virtual learning environments with students and professional trainers. *International Journal of Advanced Corporate Learning* 8:14–20.

Tranfield, D., Denyer, D., and Smart, P. 2003. Towards a methodology for developing evidence-informed management knowledge by means of systematic review. *British Journal of Management* 14:207–222.

Wollschlaeger, M., Sauter, T., and Jasperneite, J. 2017. The future of industrial communication: Automation networks in the era of the Internet of things and Industry 4.0. *IEEE Industrial Electronics Magazine* 11:17–27.

Zaske, S. 2015. Germany's vision for Industrie 4.0: The revolution will be digitized. www.zdnet.com/article/germanys-vision-for-industrie-4-0-the-revolution-will-be-digitised/ (accessed January 29, 2018).

4

Enhancing Ethical Awareness in Future Generations of Engineers

William P. Schonberg, Joel P. Dittmer, and Kevin C. Skibiski
Missouri University of Science and Technology

CONTENTS

4.1 Introduction

The first fundamental canon of the Code of Ethics of the American Society of Civil Engineers (ASCE) reads as follows: "Engineers shall hold paramount the safety, health and welfare of the public ...." Following closely behind is the sixth canon: "Engineers shall act in such a manner as to uphold and enhance the honor, integrity, and dignity of the engineering profession and shall act with zero-tolerance for bribery, fraud, and corruption" [1]. It is clear that the professional practice of civil engineering, and, in fact, all engineering, is predicated on the ethical behavior of the members of that profession.

Engineers have great power to design, test for, and make things of great value, but usually also at great cost. Because of this, we are moved to mindfully educate future engineers and endeavor to instill not only knowledge of the professional ethical codes for their specific engineering discipline but also an appreciation of various (historical) cases of ethical design and implementation failures (e.g., the Space Shuttle Challenger accident, the Bhopal accident). However, they must also be challenged to consider whether they have responsibilities outside of what is dictated by the law and their employer, which itself is usually dictated by market demand at the corporate level.

For the purposes of this chapter, we define ethics as the set of rules and principles of conduct recognized as belonging to a particular group of individuals or, in this case, a profession [2]. This differentiates ethics from morals, which can be said to characterize correct behavior, and often have a religious basis. If we expect our students to be ethical professionals when they enter the professional arena, we must provide them with a solid foundation in ethics and an understanding of the meaning of ethical behavior.

The importance of ethics education and training is also underscored by the American Accreditation Board for Engineering and Technology (ABET), which has imposed the following requirement on all programs seeking accreditation: programs must demonstrate that they instill in their graduates "an understanding of professional and ethical responsibility" [3]. In this chapter, we discuss some of the cross-disciplinary methods we use at the Missouri University of Science and Technology to engage our students to inform and educate them on proper ethical behavior. Although specific instructional engagement examples are provided from the Civil, Architectural, and Environmental (CArE) Engineering Department and the Department of Arts, Languages, and Philosophy (ALP), the techniques discussed can be applied to any number of departments and programs (e.g., business, computer science).

We also attempt to show that thinking of engineering ethics merely as a system of rules regarding how to be a "good employee" is perhaps too limiting, and that perhaps engineering as a profession might have a responsibility to grapple with what the purposes of it, as a profession, are supposed to be. We also discuss how our current approaches to ethics education and training can be expanded. In both departments, we have seen that maximizing discussion between students concerning ethical issues is a significant way of getting them to seriously consider the importance of the profession they are entering. This chapter concludes with brief discussions of several case studies involving ethics violations and how they were resolved. These case studies not only highlight the need for continuing and expanded ethics education, but they can also be used in discussions with engineering students from any discipline.

4.2 Ethics Education in the CArE Engineering Department

In the CArE Engineering Department, we discuss ethics in the sophomore-level course *Engineering Communication* in two modules: Academic Ethics and Professional Ethics. In the first module, discussions were centered on academic dishonesty—What is cheating? What is plagiarism? What is the difference? Students are presented with several scenarios, and with each scenario, there is a sequence of questions that attempt to challenge what many of them see as acceptable behavior. For example, Figure 4.1 shows a slide that forms the basis of one of the discussions.

The questions are presented one at a time. After a question is presented, the students are given a few minutes to discuss their feelings in small groups; then students are prompted to share their thoughts, and a general class-level discussion ensues. Although the early questions may lend themselves to a "right" or "wrong" answer, such certainty is rarely the case in the later questions, where ethical shades of gray might enter the discussions.

To start the discussion on plagiarism, students are given the hypothetical scenario shown in Figure 4.2. Following a discussion of this situation, students are given first the question of who here is being dishonest. Mike? Todd? Or possibly .... both? Then the students are further told that Chris, who sits nearby, saw this happening ... and the students are then asked whether or not Chris is obligated to tell the instructor what she saw. This invariably gives rise to one of the more spirited discussions of the semester.

The discussion of plagiarism concludes with some comments on how to avoid plagiarism, what is "common knowledge," and admonition to stay clear of paraphrasing if at all possible. Students are also provided with online references that reinforce the points made during the discussion.

The Professional Ethics Module begins with "from the ground up" discussion of what is engineering, what is a profession, what is a professional, and finally, what is an engineering professional. At the heart of the final construct is that an engineering professional, by definition, admits to adhering to a code of ethics applicable to the practice of engineering. The two codes of ethics appropriate for students in the department are discussed (i.e., the National Society of Professional Engineers Code of Ethics and the ASCE Code of Ethics), followed by in-class discussions of several case studies.

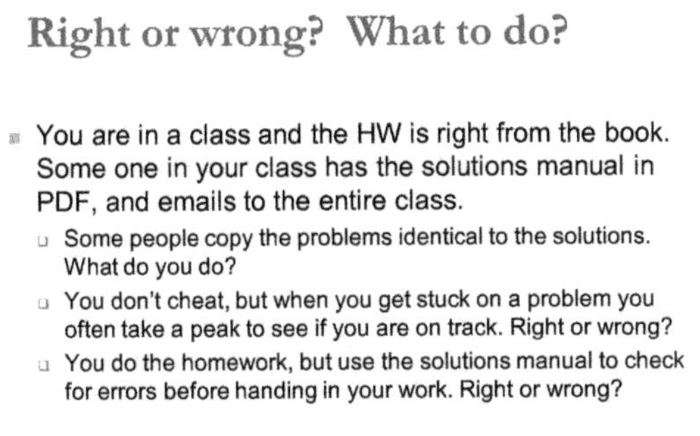

Right or wrong? What to do?

- You are in a class and the HW is right from the book. Some one in your class has the solutions manual in PDF, and emails to the entire class.
  - Some people copy the problems identical to the solutions. What do you do?
  - You don't cheat, but when you get stuck on a problem you often take a peak to see if you are on track. Right or wrong?
  - You do the homework, but use the solutions manual to check for errors before handing in your work. Right or wrong?

FIGURE 4.1
Slide from Academic Ethics Module in *Engineering Communications* course at Missouri University of Science and Technology.

What to Do?

http://www.osai.umn.edu/case.html

▪ Chris, Todd, and Mike are assigned to work on several group assignments together in a history class this semester. One of their projects include each of them researching different events on a given time line, and then combining information together.

On a test that covers some of the information gathered by Todd, Mike cannot remember what the answers are. He reasons that because the three of them had worked on the project together as a collaboration and got a good grade, it shouldn't be a problem to ask Todd what the answers are. Since they sit not far from each other in class, Mike asks Todd to tell him the answers. Todd does not want to offend his friend, so he moves his arm so Mike can see his paper.

FIGURE 4.2
Hypothetical plagiarism scenario.

Figure 4.3 shows one of the scenarios that is discussed that is also meant to lead to a discussion of the difference between "ethics" and "morals." A lively discussion typically ensues when the instructor asks whether or not it might be possible for an immoral person to be ethical.

Several additional scenarios are introduced, now having more to do with specific behaviors of engineering professionals that could be viewed as potentially unethical. The following is a list of some of the scenarios brought forth and discussed:

- A professional engineer licensed in the state of Minnesota now works in Missouri. She does not sign plans or other engineering documents with her professional seal. However, she does use the title of Professional Engineer (PE) in advertising engineering services for her company. Is this ethical?

- The owner of an engineering firm who is a PE dies suddenly. His widow takes over running the company. She is not a licensed

Example of ethics vs. morality

▪ Two lawyers heard their client confess to a crime for which someone else was convicted and was in jail. Their Code of Ethics requires client confidentiality. They did not step forward with the information that the convicted person was innocent until after their client had died.
▪ Was this ethical? Moral? What would you do? What if the "someone else" was convicted for the crime and scheduled to be executed?

FIGURE 4.3
Scenario introduced leading to a discussion of the difference between ethics and morals.

engineer. A family friend who is a PE helps out by signing plans and specs for her even though he is not employed by her company. Is this ethical?

- A contractor who has built several grocery stores in the past convinces an owner to use design/build for his new store construction. He contracts with a design firm to use the plans for a previous store with minor site-specific changes as plans for this new store. The contractor does not allow the design firm to make periodic inspections to confirm their design was being followed only allowing one inspection by the design firm at substantial completion. What in the design firm's ethical responsibility?

Concepts learned are reinforced by a writing assignment involving another case study that is graded not so much on grammar and syntax but more on clarity of thought and appropriateness of conclusions.

References to ethical decision-making are made throughout the civil engineering, architectural engineering, and environmental engineering programs of study, culminating in further, more in depth, discussions in the Senior Seminar course that is required of all majors. Statutes and rules pertaining to the practice of engineering in the State of Missouri (where the university is located) are discussed in detail, as well as concepts such as "standard of care," what exactly is a "responsible code of practice," and what someone should do if he or she witnesses wrongdoing in the workplace. The department continues to expand its instruction of ethics and the ethical practice of engineering by adding new modules and new material in its Senior Seminar course.

4.3 Ethics Education in the ALP Department

In the ALP Department, ethics (and specifically engineering ethics) are discussed in a number of courses where students consider the three major components of engineering ethics curriculum. We will give some argumentation for why there are these three of the major components, and in doing so, we will discuss the methodology by which students are made aware of these very important components. For example, not only do we discuss the general ethical principles of various engineering subdisciplines, and not only do we examine at specific engineering cases, but we also get students to engage in material which at first might not be obvious to them as being relevant to their intended professions. Again, although the concepts and discussions below are centered around engineering ethics, they can easily be applied to other departments and programs outside the field of engineering.

Additionally, recent high-profile technological accidents and failures (e.g., the Hyatt–Regency walkway collapse [4], the Challenger Space Shuttle accident [5], and the Chernobyl [6] and Bhopal [7] plant accidents, to name a few) are used to focus conversations with students on important issues. These issues include:

- The role of the engineer/manager (Does one role trump the other, and if so, which one? Are the two roles reconcilable such that there really is no conflict, or are they irreconcilable so that assigning someone to such a role will inevitably lead to ethical problems?)
- Safety and risk (Should engineers and engineering companies be concerned with producing safe products, or should they concern themselves with merely getting a product to market and have it be the role of a regulatory or safety agency be concerned with potential risks?)
- What kinds of projects engineers should participate in (Should they participate in the development of weaponry? And if so, what kind of weapon production is ethically permissible?)

Before discussing the primary and most contentious point of this section (which has to do with a more expansive way of understanding the ethical responsibilities of engineers), it is important to discuss in more detail the distinction between ethics and morality. Doing so will provide more legitimacy to this distinction, as well as highlighting just how important is engineering ethics. Finally, it will provide a nice segue to the main point of this section, reinforcing the idea that we should be careful not to thoughtlessly import ethical intuitions from one domain of practices (e.g., familial life) to another, like engineering.

There are various ways to understand the difference between ethics and morality. One way is to think of ethics as the study of morality, just as chemistry, for example, is the study of chemical structures and processes. It is interesting to note that the 19th-century philosopher, Hegel, made a somewhat similar distinction between ethics and morality. For him, morality was thought of as a more abstract way of guiding one's life, whereas ethics was a more concrete approach. Moral issues are going to be more abstract and not contextualized to political, social, familial, and other institutional considerations. Ethical issues, on the other hand, are contextualized to these considerations. Additionally, the distinction we make between ethics and morality, just like Hegel's distinction, highlights the importance of how normative issues (What should I do? What do I value? What kind of person do I want to be?) are not considered from the perspective of decontextualized persons, empty of the specific roles, relationships, and indeed, professions that they take on. One can, therefore, see how from the perspective of a more abstract approach, the very same person can do something wrong, which from the perspective of the specific role they take on is actually ethically appropriate, perhaps even required.

As an example to help reinforce the importance of distinguishing ethics and morality as we have articulated it herein (or more correctly, the importance of making the distinction in which we have used "ethics" and "morality" as the terms used to refer to the things being distinguished)—this is an example often used while teaching both business ethics and engineering ethics—suppose that your firm is making a bid for a specific project, say the design of a new hospital building. You also know that the dates set for design will take place 3 months from now; that is, whichever firm gets the design project, none of them will start the design for another 3 months. One possible problem, though, is that you have two chief engineers who will retire within a few months. Nevertheless, you will be recruiting some experienced engineers in the meantime, and additionally, you still have a number of experienced engineers on staff. As you prepare your bid for this project, you decide to not disclose this upcoming change in personnel in your bid papers.

Although outright deception is likely to be considered to be morally/ ethically forbidden, it is not as clear (i) whether failure to disclose this information would, in fact, be considered as deception and (ii) whether, even if deceptive, it is morally/ethically wrong. However, once we employ the morality versus ethics distinction note previously, we can say something like this: whereas failure to disclose the upcoming personnel change would likely be considered morally wrong (as this comes close to deception, and deception is morally wrong), it is not ethically wrong relative to currently accepted practices of how bids are made in industry. Furthermore, we only think that it is immoral in an abstract and universal way because we have abstracted this thought from our own experiences where failure to disclose would be wrong (e.g., sharing with your spouse some important informa- tion concerning a refinancing your jointly held mortgage that you intend to undertake). Of course, in this spousal example, failure to disclose would also likely be considered unethical as well.

And so, finally, we can directly discuss the main, and possibly more contentious, point of this section, which is that engineers may have certain ethical responsibilities that extend beyond being merely a "good employee" or even making sure not to do harm. Notice earlier with the morality versus ethics distinction, engineers may be "let off the hook" ethically once we understand the context in which engineers practice their profession. But similarly, they may be ethically required to do things that when considered from an intuitive perspective, or from an abstract perspective, might not be obvious.

We can motivate this idea further with the case called Shallow Pond [8], slightly modified. Suppose that upon walking to class, you see a child drowning in a shallow pond. No one else is around to help; it's only you. And although you eventually need to get to class, there is no pressing need to get to class for a while. Additionally, you aren't wearing very expensive clothes. Finally, saving the child would pose no serious threat to yourself. Suppose then that you still decide not to help and save the child. Most people would say that you've done something seriously morally (and ethically) wrong. Part of the explanation of why you've done something wrong is that you've allowed someone to die when (i) you had plenty of capacity to prevent that from happening, and (ii) in doing so, you would not have brought serious harm to yourself or at least none in any significant or lasting way.

Similarly, engineers may be in a position to prevent the suffering of others because of the capacities they have without making themselves significantly worse off, if at all. Given that engineers have unique knowledge and skills for designing and building products and systems that can benefit many people, especially those who are the least advantaged in the world, there is a strong case for engineers being ethically *required* to help such people in much the same way that any able-bodied person would be ethically required in Shallow Pond.

Of course, it is incumbent upon us as faculty to provide opportunities for students to learn how to engage with the possibilities of improving the lives of others, especially those who are least advantaged. One manner in which to accomplish this would be to introduce more anthropology into our courses or at least knowledge acquired by anthropologists. This would provide a framework within which students can begin to understand the values and commitments, as well as legitimate, concrete needs, of the group of people they are attempting to help. Why is this important? First, what might be considered necessary for one group might prove unnecessary for another. Second, one could imagine a group of engineers making a number of proposals for some solution to a group's problem. But choosing a proposed solution without looking at whether the proposal coheres with the group's values and commitments is ethically suspect. If a proposed solution would conflict with the group's values or cause for the compromising of their values, then this proposed solution would be problematic. Alternative proposals then should be made and with the ultimate selection based on whether they promote, and at least do not conflict with, the values and commitments of the group.

One might suspect that the considerations in the previous paragraphs lend themselves to a kind of ethical relativism. One very contemporary kind of relativism can be found in Ref. [9]. And with such a relativism, where what's ethically right varies according to culture, one would probably be correct that any particular professional engineering aiming for universality is unachievable. First of all, though, the considerations of the previous paragraph don't necessarily entail such a strong relativism. Second, even if a weak form of relativism were the case—where what is valued varies from one group or culture—this does not mean some very basic values can't be shared, and shared universally. Basic values such as protection from harm, accordance to veracity, and fairness are some possible candidates for (at least) widely shared values among radically divergent groups and cultures. Thankfully, though, most (if not all) engineering ethics codes acknowledge the role of engineers in, if not in enhancing these values, at least respecting them.

So, it might turn out that these engineering codes have universal applicability. It's just that what is needed is a further understanding of what constitutes a harm for a particular group/culture, or what counts as honest disclosure opposed to deception for a particular group/culture, or finally what amounts to a fair deal for a particular group/culture.

To conclude this section, we reemphasize the importance of getting future engineers to be mindful of the three components discussed earlier. Furthermore, although there are multiple ways of distinguishing ethics and morality, one such way is particularly useful in getting future engineers to think about how role and context specific their ethical responsibilities are. Furthermore, in virtue of their particular abilities as engineers, they may be ethically required to aid (to make things better, not just not-worse) in ways that non-engineers are not so required. Finally, in being sensitive to the differing values of groups and cultures, professional engineering ethics codes are not rendered misguided or unachievable (as they aspire to universality). Instead, it may be that there are some basic shared values between cultures that are manifested in importantly different ways, and yet, engineering ethics codes account for those basic values. One thing for educators to do is to make students sensitive to this fact and to guide them in acquiring knowledge concerning these cross-cultural disciplines.

4.4 Some Case Studies for Discussion

The following paragraphs discuss real cases (with names omitted) that were brought before and adjudicated by the State of Missouri's Board for Architects, Professional Engineers, Professional Land Surveyors, and Professional Landscape Architects (hereafter referred to simply as, "the Board"). They highlight the need for continuing our efforts in educating future engineers to be forever vigilant against lapses in ethical judgment. Although centered around engineering activities related to public works, they can be packaged together to be part of a multidisciplinary discussion of ethics and ethical decision-making.

4.4.1 Changing Jobs

A young Professional Engineer worked for a mid-sized consulting firm in Missouri. One of his firm's clients was a Public Water System District (PWSD). After a few years, the young Professional Engineer felt the urge to go into business for himself, so he resigned and started a small consulting firm. PWSD liked the work the young Professional Engineer had done for them over the years. They terminated the contract with the mid-sized consulting firm and hired the young Professional Engineer as an independent consultant.

The mid-sized consulting firm filed a complaint with the Board against the young Professional Engineer. Key elements of the complaint included misconduct, violation of professional trust, conflict of interest, and compensation from more than one party for services pertaining to the same project.

This may be looked at as a possible violation of the ASCE Code of Ethics Fundamental Principle 2 and Fundamental Canon 4, which says "Engineers uphold and advance the integrity, honor and dignity of the engineering profession by being honest and impartial and serving with fidelity the public, their employers and clients.

Engineers shall avoid all known or potential conflicts of interest with their employers or clients and shall promptly inform their employers or clients of any business association, interests, or circumstances which could influence their judgment or the quality of their services. Engineers shall not accept compensation from more than one party for services on the same project, or for services pertaining to the same project, unless the circumstances are fully disclosed to and agreed to, by all interested parties."

The issue the Board focused upon was the intent of the young Professional Engineer when he resigned his position with the mid-sized consulting firm and the timing of the events that occurred. It was felt that the young Professional Engineer had made a mistake in his decision-making, but it had not affected the health, safety, or welfare of the public with his actions. He was given a public censure and asked to successfully complete an 8-h online ethics course from the National Institute for Engineering Ethics at the Murdough Center for Engineering on the campus of Texas Tech University.

4.4.2 Bribery

During a normal license renewal, an out-of-state Professional Engineer for a Public Works Department in a large mid-western city notified the Board he had pled guilty to a federal offense involving bribery. He was sentenced to 44 months in federal prison and ordered to pay $134,000 resolution of funds to the city.

This is a violation of the ASCE Code of Ethics Fundamental Principle 2 and Fundamental Canon 6, which says "Engineers uphold and advance the integrity, honor and dignity of the engineering profession by being honest and impartial and serving with fidelity the public, their employers and clients.

Engineers shall not knowingly engage in business or professional practices of a fraudulent, dishonest or unethical nature. Engineers shall be scrupulously honest in their control and spending of monies, and promote effective use of resources through open, honest and impartial service with fidelity to the public, employers, associates and clients. Engineers shall act with zero-tolerance for bribery, fraud, and corruption in all engineering or construction activities in which they are engaged. Engineers should strive for transparency in the procurement and execution of projects. Transparency includes disclosure of names, addresses, purposes, and fees or commissions paid for all agents facilitating projects. Engineers should encourage the use of certifications specifying zero-tolerance for bribery, fraud, and corruption in all contracts."

The Professional Engineer took the correct ethical action after his sentence in notifying the Board of the offense; however, his lapse in moral judgment caused his PE license to be revoked. After his sentence is served and restoration of funds complete, he may, if he wishes, apply to have his PE license reinstated. There is no guarantee the sitting Board at that time will consider reinstatement.

4.4.3 Plagiarism

A mid-sized consulting engineering firm in Missouri issued construction plans in 2012 for a new medical center in rural Missouri. In 2013, the new medical center administrator requested copies of the building drawings, which were sent electronically in pdf format. The administrator came back and requested AutoCAD drawings, which were issued to him after he signed a release form stating "the documents are the property of the engineer and would not be reused for any purpose other than the original project."

A small consulting engineering firm in Missouri issued plans for another new rural medical center in November 2013. It appeared that the contents of the new rural medical center plans issued by the small consulting firm were copied directly from the AutoCAD files produced by the mid-sized firm for the first medical center.

The mid-sized consulting firm filed a complaint with the Board against the small firm, noting that the project they completed in 2012 was designed in 2009 and followed the 2003 building codes, while the project done by the small consulting firm was to follow the 2009 building code, but the details were copied and had not been changed to reflect changes in the building codes.

This could be a violation of the ASCE Code of Ethics Fundamental Principles 1 and 2 and Fundamental Canons 1, 5, and 6, which says "Engineers uphold and advance the integrity, honor and dignity of the engineering profession by using their knowledge and skill for the enhancement of human welfare and the environment, and being honest and impartial and serving with fidelity the public, their employers and clients.

Engineers shall recognize that the lives, safety, health and welfare of the general public are dependent upon engineering judgments, decisions and practices incorporated into structures, machines, products, processes and devices. Engineers shall approve or seal only those design documents, reviewed or prepared by them, which are determined to be safe for public health and welfare in conformity with accepted engineering standards. Engineers shall give proper credit for engineering work to those to whom credit is due, and shall recognize the proprietary interests of others. Whenever possible, they shall name the person or persons who may be responsible for designs, inventions, writings or other accomplishments. Engineers shall not knowingly engage in business or professional practices of a fraudulent, dishonest or unethical nature."

Representatives of the small consulting firm met with the Board, after which the Board obtained additional information which led the Board to believe the representatives had not been truthful in their testimony to the Board. The PE licenses and firm Certificate of Authority were placed on probation.

4.4.4 Profit over Service to the Public Health, Safety, and Welfare

A central Missouri city building official filed a complaint against a PE (Chemical) alleging incompetence, misconduct, assisting an unlicensed person to practice engineering, practice of architecture, practice outside his area of expertise, and breach of professional trust. A set of building plans (architectural, structural, mechanical HVAC, plumbing, and electrical) were submitted, signed, and sealed by the PE (Chemical).

This could be a violation of the ASCE Code of Ethics Fundamental Principle 1 and Fundamental Canon 2, which says "Engineers uphold and advance the integrity, honor and dignity of the engineering profession by using their knowledge and skill for the enhancement of human welfare and the environment.

Engineers shall undertake to perform engineering assignments only when qualified by education or experience in the technical field of engineering involved. Engineers may accept an assignment requiring education or experience outside of their own fields of competence, provided their services are restricted to those phases of the project in which they are qualified. All other phases of such project shall be performed by qualified associates, consultants, or employees. Engineers shall not affix their signatures or seals to any engineering plan or document dealing with subject matter in which they lack competence by virtue of education or experience or to any such plan or document not reviewed or prepared under their supervisory control."

At a hearing before the Board testimony was given that indicated that PE (Chemical) did not have education, training, or experience in designing buildings. The outcome was a 3-year probation of his PE license, including quarterly reports to the Board for projects on which he was working. One such quarterly report listed three building design projects for which he signed and sealed all of the plans. PE (Chemical) openly and arrogantly ignored his first violation and probation restrictions and continued to practice outside his area of expertise.

PE (Chemical) was again called before the Board, this time to explain how he was now qualified to do this work. In lieu of going through a probation violation hearing, the PE (Chemical) voluntarily surrendered his PE license and is no longer practicing.

4.5 Concluding Comments

We hope that in this chapter we have helped illuminate some of the inter-disciplinary approaches taken in educating Missouri S&T engineering students in ethics. As should be hopefully apparent, we get students to explore, and actually do, ethics through the consideration of a number of case studies and examples. In our chapter, we have made a distinction between morality and ethics, and emphasized the importance of looking at issues in not just a moral way but also an ethical way. Once again, the ethical approach is one that examines issues in light of the contextual features of the institution under which they take place. Finally, we encourage students to think of the possibility of being a "good engineer" is not fully exhausted by simply being a "good employee." Instead, it may be the case that because of the unique position engineers are in whereby they are able to help the plight and problems of others, engineers are then ethically required to align and focus their work to, in fact, help others. We hope to instill this responsibility in our students by making our courses more "anthropological," in that we hope to educate students that they must be aware of what various groups, in fact, actually need and desire, and what they, in fact, actually value.

References

1. ASCE, *Code of Ethics.* www.asce.org/Ethics/Code-of-Ethics/.
2. Gayton, C. *Legal Aspects of Engineering*, 9th Edition. Kendall Hunt, Dubuque, IA, 2012.
3. ABET, EAC Criteria 2013–2014 (Criterion 3f). http://abet.org/uploadedFiles/ Accreditation/Accreditation_Step-by-Step/Accreditation_Documents/ Current/2013-2014/eac-criteria-2013-2014.pdf.
4. Marshall, R.D., Pfrang, E.O., Leyendecker, E.V., Woodward, K.A., Reed, R.P., Kasen, M.B., and Shives, T.R. Investigation of the Kansas City Hyatt Regency Walkways Collapse. Report No. NBS BSS 143, National Institute of Standards and Technology, Washington, DC, May 31, 1982.

5. Report to the President by the Presidential Commission on the Space Shuttle Challenger Accident. Washington, DC, June 6, 1986.
6. The Chernobyl Accident: Updating of INSAG-1. Safety Series Report No. 75-INSAG-7, International Nuclear Safety Advisory Group, International Atomic Energy Agency, Vienna, 1992.
7. Browning, J.B. *Union Carbide: Disaster at Bhopal.* Union Carbide, India, 1993.
8. Unger, P. *Living High and Letting Die.* Oxford University Press, Oxford, 1996.
9. Prinz, J. *The Emotional Construction of Morals.* Oxford University Press, Oxford, 2009.

5

Merging Literature and Voices from the Field: Women in Industrial and Systems Engineering Reflect on Choice, Persistence, and Outlook in Engineering

Federica Robinson-Bryant
Embry-Riddle Aeronautical University-Worldwide

Alice Squires
Washington State University

Gina Guillaume-Joseph
MITRE Corporation

Angela D. Robinson
International Council on Systems Engineering

Shanon Wooden
University of Central Florida

CONTENTS

5.1 Introduction

In a time when systems are increasingly complex, the need for a more diverse and productive engineering labor force continues to gain traction. Less than 16% of seats on corporate boards are held by women (GAO, 2016), and the notion of a "boys club" is often realized in many science, technology, engineering, and mathematics (STEM)-based organizations. Significant legislation and initiatives have aimed to increase the participation of underrepresented groups like women in STEM with mixed results.

In general, women have been cited as less likely to enter STEM fields and more likely to leave STEM careers across many regions in the world (UNESCO, 2017). Still, less than one-third of science and engineering professionals are women, and the ratio is substantially less for the engineering field. These disparities proliferate into engineering in such a way that challenges women's choice to pursue and persist in the field. In *The Only Woman in the Room: Why Science Is Still a Boy's Club* (2016), Eileen Pollack claimed, "I didn't stand up to my jailers because I realized how quickly I would have been overpowered." This metaphor implies the uphill journey confronting women in engineering as they face individual, social, organizational, and societal barriers throughout their career.

Consequently, the goal of gender parity in engineering presents itself as a myth that requires explicit and intentional understanding of the problem in order to ignite transformative change across the discipline. The implication is that the next few generations may not see this parity achieved so mind-set shifts and identification of support mechanisms become an interim strategy.

This chapter attempts to isolate the topic of gender in the professional engineering experience, with an emphasis on illuminating multiple perspectives on factors influencing the choice to pursue and persist in industrial and systems engineering (ISE) while understanding women's outlook on success as an engineer. Past theoretical and empirical studies are vast, but its prevailing focus on the undergraduate students' experience and decisions to persist in academic programs leave much room for conversation around women's experiences after graduation.

It is well evidenced that the professional engineering experience will vary from person to person, based on the intersection of an array of characteristics such as gender, race, culture, disability, sexuality, and age, their interactions with each other and external influences. Yet researchers have found that certain factors and patterns exist. This chapter will strive to uncover what research has posited to date, while also presenting the prompted perspectives of ISE women from various career paths and experience levels using qualitative methods. Admittedly, this is not an attempt to generalize the issue or suggest a solution but rather to inform the ongoing conversation and shed light on support resources to increase the persistence of women in ISE.

5.2 A History of Women in Engineering

Individuals have been performing engineering tasks as early as the beginning of time, based on Merriam Webster's definition of engineering as "the application of science and mathematics by which the properties of matter and the sources of energy in nature are made useful to people," but the first known use of the term was in 1697 (Webster, 2018). People, both women and men, worked in roles deemed as trades until earning recognition as a profession in the scientific community in 1916 (Tietjen, 2017). This professionalization of engineering caused further barriers for women because it aimed to distinguish between engineers with formal education and experience and those engineers that learned on the job. The limited number of engineering schools that accepted women made it more difficult for women to earn formal credentials. Not unlike today, white males dominated engineering and those women who chose to pursue engineering were often met with resistance and oppression in their pursuit of an engineering identity.

The 1890 U.S. Census marked the first time females were captured as occupying an engineering role although the notion of "woman" did vary from today's use of the term. The report considered females over the age of 10 years old and showed 124 females occupying an engineering role compared to 43,115 males in the same age range (<1%). At this time, the engineering classification was termed "Engineering (civil, mechanical, electrical and mining) and surveyors" and featured limited lineage to the plethora of domains of present times. The 1890 Census Occupation Codes included **1950**-civil engineers; surveyors, **1860**-civil and mechanical engineers; electricians, **1870**-engineers (civil; land surveyors), and **1180**-engineers (civil) under the engineering field.

The first documented woman graduate from an academic, engineering program was Elizabeth Bragg in 1876 (Tietjen, 2017). However, there are records of female students matriculating in engineering programs through the doctorate level as "special students" prior to this time. This denotation meant that they would never be documented as engineering graduates despite having completed the coursework.

More than a century later, the disparity in representation persists. The 1990 U.S. Census captures much progress in the field with clear delimiters among surveyors, architects, and engineers that were not present in 1890 and 13 distinct disciplines of engineering. Much growth in the representation of women in the field is also present, as women accounted for approximately 9% of the engineering workforce. Figure 5.1 captures the gendered representation

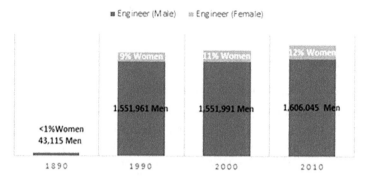

FIGURE 5.1
The U.S. gender representation in engineering from 1890 to 2010.

in engineering in the United States from 1890 to 2010 (U.S. Census, 1890, 1990, 2000, 2010). This representation in the labor force remains somewhat stagnant to date despite heavy investments in academic, professional, and federal interventions by organizations like the National Science Foundation. In 2016, federal agencies reported investing $2.9 billion across 163 STEM education programs with about 76% of these programs tracking female and women participants and were primarily intended to serve underrepresented groups (GAO, 2018).

Conversely, the gender gap among the total U.S. labor force continues to narrow; such trend does not yet materialize in engineering. Women served as almost 17% of the total labor force in 1890. There has been a significant increase in female participation in the workforce over the past 30 years, specifically 46% representation in 1990 and 47% representation in 2000 and 2010. This shift to near parity is likely a result of the social and political changes in the role of women in U.S. society. For example, in earlier years, women were expected to get married, have children, and stay home (Tietjen, 2017), which was the choice of many like the first engineering graduate who never actually entered the engineering profession. However, more and more women today assume dual roles to balance both home life and a career or hold reverse roles with their spouse or support systems (Figure 5.2).

5.3 Women in ISE

In May 2015, there were about 1.6 million engineers employed in the United States. Mechanical, civil, and industrial engineers together made up half of all employed engineers (Bureau of Labor Statistics, 2016). Industrial engineering (IE) is a term used as early as 1911 (Webster, 2018) and is often characterized

U.S. LABOR FORCE OVER TIME

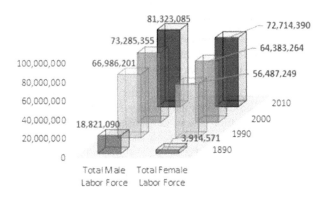

FIGURE 5.2
The U.S. labor force from 1890 to 2010.

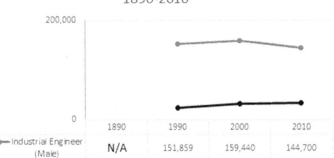

FIGURE 5.3
Total engineers in IE profession, 1890–2010.

with the mantra that "IEs make things better" and are the "jack of all trades." The field deals heavily with the intersection of people, materials, and energy over the life cycle of systems. Notably, Figure 5.3 shows that there is an increasing trend in women participation in IE, and IEs account for about 10% of all engineers (U.S. Census, 2010). Women accounted for about 14% of the IE labor force in 1990, 17% in 2000, and 19% in 2010. The slightly higher rate of representation in this discipline and the very nature of IE work support the notion of IE work being more attractive to women (Brawner, Camacho, Lord, Long & Ohland, 2012). However, one contributor cautions, "we need to move forward with more female scientists in all fields rather than relegate them to certain subspecialties..." (Nilsson, 2015).

The U.S. Census nor the Bureau of Labor Statistics include the discipline of Systems Engineering in reported classification standards. The often debated question of whether systems engineering is its own discipline could be just cause. The International Council on Systems Engineering (INCOSE) defines systems engineering as an approach that

> integrates all the disciplines and specialty groups into a team effort forming a structured development process that proceeds from concept to production to operation...[and] considers both the business and the technical needs of all customers with the goal of providing a quality product that meets the user needs.

Findings from the *Atlas Report* confirm the complexity of the systems engineering identity (SERC, 2018). Many SEs earn an undergraduate degree in a traditional engineering domain and may identify as such, even after assuming a systems engineering role. In addition, many academic institutions have historically blended ISE programs or typically do not offer systems engineering degrees at the undergraduate level. The report defines many roles a systems engineer may assume via the Atlas Roles Framework, while admitting these roles may not be complemented by a formal SE title.

Similarly, the varied nature of ISE roles (SERC, 2018) and the historical context dilute discipline categorization for many pioneers in ISE. Nonetheless, their influence has clear traceability to the field. The contributions of the women captured in Table 5.1 have paved the way for the wave of women engineers that would come after them. Their achievements, despite the climate, serve as inspiration and motivation to persist.

Over the years, many institutions have managed to increase the recruitment and retention of women in engineering academic programs in an effort to reach gender parity in engineering (Leslie, McClure & Oaxaca, 1998; Servon & Visser, 2011). Figure 5.4 shows the percentage of engineering, IE, and systems engineering degrees earned by women during 2006–2016 at the undergraduate and graduate levels based on the Department of Education's Classification for Instructional Programs (National Center for Education Statistics Digest: Tables 290.2009, 317.2010, 318.30.2011–2016).

However, a comparison of Figure 5.4 to Figures 5.1 and 5.3 captures a slightly different outcome. Engineering is one of the most segregated professional occupations in the United States (Cech, Rubineau, Silbey & Seron, 2011; Singh et al., 2013). While the number of degrees awarded in engineering, IE, and systems engineering continues to grow in many cases, the number of women working in engineering remains mostly unchanged. Women account for more than 20% of engineering school graduates, but between 10% and 11% of practicing engineers are women (Buse, Perelli & Bilimoria, 2009; Bureau of Labor Statistics, 2016; Fouad, Singh, Cappaert, Chang & Wan, 2016).

TABLE 5.1

Inspirational Women in ISE

Dr. Leslie Ann Benmark (industrial engineering)	1944–	International expert in engineering education and accreditation; past president of ABET; elected member of the National Academy of Engineering (NAE)
Dr. Mary (Missy) Cummings (systems engineering)	1966–	One of the United States Navy's first female fighter pilots; author; successful professor
Dr. Nancy Jane Currie-Gregg (industrial engineering)	1958–	U.S. military Colonel and NASA astronaut with over 1,000 h of spaceflight
Dr. Lillian Moller Gilbreth (industrial engineering)	1878–1972	One of the first female engineers to earn a PhD; considered to be the first industrial/organizational psychologist; an early pioneer in applying psychology to time-and-motion studies; the first female to be inducted into NAE
Linda Parker Hudson (industrial and systems engineering)	1950–	First female CEO in the defense industry; former CEO and COO of BAE Systems
Dr. Nancy Leveson (systems engineering)	1944–	An elected member of the NAE; a leading American expert in system and software safety
Ellen Henrietta Swallow Richards (industrial engineering)	1842–1911	Founder of the home economics movement characterized; first to apply chemistry to the study of nutrition; first woman in America accepted to any school of science and technology
Virginia Marie "Ginni" Rometty (systems engineering)	1957–	First woman president and CEO of IBM; recipient of many prestigious awards and rankings
Dr. Laura Tremosa (industrial engineering)	1937–	First Catalan woman, and the second Spanish woman to graduate in industrial engineering; influential feminist

Fouad et al. (2016) found that almost half of women who graduate from engineering decide to leave an engineering career but half of those remain in the engineering profession. This is compared to a 10% departure rate from engineering for men. Frehill (2008) also found that more than two-thirds of women leave engineering by 15 years post-degree, double the rate found for men. This rate is further reduced to 7 years for women engineering faculty in academia (National Academies Press, 2006).

A much-anticipated shift lies ahead for engineering and engineering education that may foster the shrinkage of the gender gap over time and the nurturing of other knowledge, skills, and abilities that interact to positively influence the engineering experience.

Engineering education, like professional engineering, has historically exacerbated the discourse experienced by women after graduation by

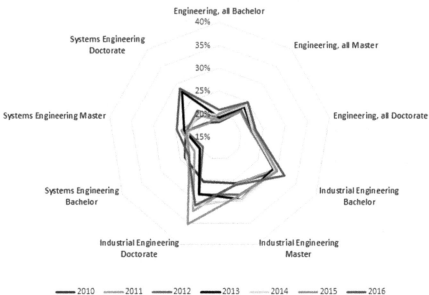

FIGURE 5.4
Percent of U.S. degrees in engineering awarded to women each year.

providing limited opportunities to work in teams that assimilate those often encountered in the engineering workplace. Accreditation Board for Engineering and Technology (ABET), first formed in 1932 as the Engineering Council for Professional Development (ECPD), is an accreditation body that defines the criteria by which programs should meet in order to be identified as an ABET-accredited engineering program. The approved changes to the 2019–2020 ABET Engineering Accreditation Committee (EAC) accreditation criteria will challenge engineering programs to more explicitly address diversity in teams and globalization in the student's academic experience.

In addition, several ongoing initiatives promote the integration of systems thinking and systems engineering in all engineering programs irrespective of domain (Adcock, Squires, Gannon & Virani, 2015; Robinson-Bryant & Norman, 2017). The value of developing systems thinking skills like Self-Awareness and Multiple Perspectives across the engineering workforce will help men and women to assume a more equitable stake in the effort to change the "hostile" or "chilly" climate women face (National Academies Press, 2006; Buse et al., 2009). Together, engineering graduates will be predisposed to many of the challenges of this dynamic earlier in their engineering experience and better positioned to support and persist.

5.4 Research on Women in Engineering

Decades of academic research, federal funding and employee interventions have been poised to address the gender gap in engineering (Fouad & Singh, 2011; Kelley & Bryan, 2018) but the impact on representation in the labor force has been relatively null. The dire needs identified in NSF's *The State of Academic Sciences* report (1989) challenged the United States to significantly increase the engineering labor force to meet the anticipated needs. Sources suggest that females and racial minorities became the source of prospective "bodies" rather than the conscious "valuing of diversity and equitable representation" (Garbee, 2017). This has led to an emphasis on getting women into engineering initiatives and programs, or the "pipeline", with an expectation that many women will not make it for a number of reasons (the "leaky pipeline" concept, see Figure 5.5). Duncan & Zeng (2005) express their discontent in pushing women into engineering "without consideration of the probability of their completion and long-term service to the engineering community." Arguably, both the causes and residual effects of women leaving engineering during an academic program or leaving engineering at some point postgraduation is multifaceted across individual, organizational, cultural, economic, political, and societal lens, and deserves more intentional consideration.

Despite competing schools of thought in the context of engineering, engineering preparation, and talent acquisition/retention (Figure 5.5), much attention has been given to how to "plug the holes" in the pipeline during academic matriculation to graduate more engineers (NSF, 1989; Atman et al., 2008; Garbee, 2017). Less common is the examination of the broader and systemic changes required to support women's persistence in their engineering careers appear.

Ultimately, the multidimensional premise that precedes the resolution to gender parity in engineering requires careful consideration of factors affecting the experience of women at multiple levels of impact. Most of what is perceived or experienced is the result of systemic and highly interrelated interactions and dependencies. For instance, what appears to be organizational discourse permeates at the social and individual levels.

5.4.1 Social Cognitive Career Theory

One of the most commonly cited theories in the work on women in engineering is the social cognitive career theory (SCCT) developed by Lent, Brown & Hackett (1994) and adapted by many researchers (Lent, Brown & Hackett, 2000; Tharp, 2002; Inda, Rodriguez & Pena, 2013; Fouad et al., 2016; Cadaret, Hartung, Subich & Weigold, 2017). The essence of the theory is that multiple individual, contextual and environmental (objective and perceived) variables interact in a transactional and reciprocal manner to frame a person's career interests, choices and performance. *Person inputs* are those inherent attributes often considered beyond the control of the individual and are often used to

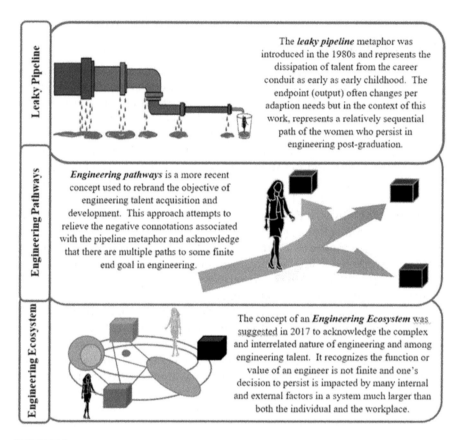

Leaky Pipeline

The *leaky pipeline* metaphor was introduced in the 1980s and represents the dissipation of talent from the career conduit as early as early childhood. The endpoint (output) often changes per adaption needs but in the context of this work, represents a relatively sequential path of the women who persist in engineering post-graduation.

Engineering Pathways

Engineering pathways is a more recent concept used to rebrand the objective of engineering talent acquisition and development. This approach attempts to relieve the negative connotations associated with the pipeline metaphor and acknowledge that there are multiple paths to some finite end goal in engineering.

Engineering Ecosystem

The concept of an *Engineering Ecosystem* was suggested in 2017 to acknowledge the complex and interrelated nature of engineering and among engineering talent. It recognizes the function or value of an engineer is not finite and one's decision to persist is impacted by many internal and external factors in a system much larger than both the individual and the workplace.

FIGURE 5.5
Diverse perspectives of engineering talent acquisition and retention.

define an individual's identity. This includes predispositions, gender, race/ethnicity, disabilities/health and other characteristics like sexuality.

These factors directly inform an individual's *learning experiences, environmental factors* from their background and the supports and barriers experienced during decision-making (or *proximal environmental influences*). The theory further suggests that resulting learning experiences directly influence one's *self-efficacy* (or belief that they can be successful) and *outcome expectations*. The belief in oneself will directly influence expected outcomes.

There are additional variables captured in the theory as influencing individual career decisions (or *choice actions*) such as *interests* and *choice goals*. These actions ultimately lead to some level of performance and attainment (positive or negative) which serves as feedback to the learning experience. The dotted paths in Figure 5.6 include moderator effects on interest-goal and goal-action behaviors.

Yet, the career choice decision is not likely a single event but rather an evolving process with multiple influences and choice points (Singh et al., 2013).

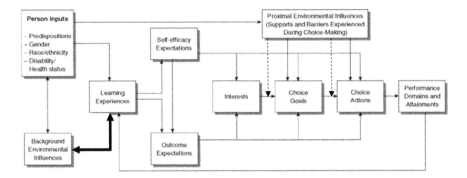

FIGURE 5.6
Adaptation of social cognitive career theory.

This perspective supports the need to extend the SCCT model to account for the impact of the learning experiences among other factors (Figure 5.6). The adapted SCCT model in Figure 5.6 adds feedback to a person's background and environmental influences due to actualized or perception changes over time. The substitution of an emboldened, bidirectional arrow extending from learning experiences suggests this change.

5.4.2 Factors Impacting Women's Decision to Persist

Researchers have studied the individual, contextual and organizational factors influencing women's decision to persist in their engineering careers (Buse et al., 2009; Buse, Bilimoria & Perelli, 2013; Kranov, DeBoer & Abu-Lail, 2014). External factors have been found to impact women's academic and professional attainment more than men (Tharp, 2002). The Project on Women Engineers' Retention (POWER) published an extensive study of over 3,700 women from an alumni pool at more than 200 universities. It found that the workplace climate was a strong factor in participants' decision not to enter the engineering profession after college or to leave the engineering profession (Buse et al., 2009).

Engineering has a reputation as a man's occupation so it is no surprise that a man's decision to assume an engineering role is viewed as an organic or authentic choice, i.e., gender authenticity (Saavedra, Araujo, Manuel de Oliveira & Stephens, 2014). However, one contributor recognized that many women find engineering interesting as well, making engineering an authentic choice for women too (Nilsson, 2015).

More specifically, researchers on this topic cite the ability to adapt to a male-dominated workplace (Buse et al., 2009) and masculine social constructs as a very relevant factor in the decision to persist (Buse et al., 2013; Servon & Visser, 2011; Saavedra et al., 2014). This belief places the burden on women to "fit", "suppress", and "emulate" in order to avoid disrupting the existing culture. Hahn & Lynn (2017) brings attention to two common perspectives that characterize interventions that promote attracting and retaining women in engineering: (i) fix the woman so that she can navigate the work environment or (ii) fix the culture of the organization so that everyone is comfortable.

The term impression management from the 1950s was referenced by Servon & Visser (2011) and refers to "the potential for individuals to adopt certain characteristics in order to create an impression of themselves, which they hope will elicit a desired reaction in another individual." The authors suggest that this entails adopting a masculine identity and behaviors like competitiveness among others in the workplace, especially toward other women and promoting higher structures of hierarchy to give the impression of conformity to the occupational culture and values.

Adopting this approach is not without an impact on the individual, her relationships and the workplace- whether negative, positive or near negligible. "Navigating the workplace" requires a level of conformity and adaption that could foster isolation and marginalization. However, research shows two aggregate outcomes that emit a glass half empty versus a glass half full effect. Participants in Servon & Visser's study (2011) viewed navigating the workplace as either a challenge or a unique developmental opportunity.

Hence, the common fallacy of integrating women into the existing system through changing their dispositions and forcing them to fit is inequitable. It fails to force organizations to accept their responsibility to create a culture that provides clear paths to advancement, supports work-life balance, eliminates earning differentials, values women's contributions and offers mentoring opportunities (Fouad & Singh, 2011; National Academies Press, 2014). A similar phenomenon occurs with other person inputs like race (e.g., systemic racism) and sexuality (e.g., U.S. military's don't ask, don't tell policy), where certain groups are marginalized and must conform to the norms of the dominant culture.

Borrowing from biology, the phenomenon of chameleons changing colors based on external stimuli and their environment creates a similar parallel. This tactic is different from what occurs in other color changing reptiles because chameleons make structural changes that rapidly adjust a layer of skin cells under their skin based on a stimulus (Teyssier, Saenko, van der Marel & Milinkovitch, 2015).

An alternative metaphor offered by Saavedra et al. (2014) associated the conditions expressed by the women in their study to the Panopticon architectural form published by Jeremy Bentham in 1791, a reasonable precept to relevant social theories. Participants in the study resonated with the notion of being stuck in the idealized prison with transparent and unbreakable glass walls that posed risks to their personal and relational well-being.

The Panopticon's conceptualization was the impetus for the social theory of panopticism. It was developed by Michael Foucault in 1975 and reflects the socially constructive nature of power structures, identity and other social constructs (Foucault, 1995, reprint). From this view, women's emulation of a masculine presence/behavior or perception of the need to do so fosters further oppression and objectification on their behalf.

A quote by Foucault (1995, reprint) offer a profound view that is not only indicative of gender in engineering but of many underrepresented attributes found among engineers:

> He who is subjected to a field of visibility, and who knows it, assumes responsibility for the constraints of power; he makes them play spontaneously upon himself; he inscribes in himself the power relation in which he simultaneously plays both roles; he becomes the principle of his own subjection.

This theory illuminates the potential impact of recognizing the disparities for women in engineering and its impact on internal, external and behavioral factors. Similarly, Cadaret, et al. (2017) found the existence of negative relationships between consciousness of discrimination due to group identity and self-efficacy.

Much of this discussion so far has focused on contextual factors but many factors are studied throughout the literature. Buse et al. (2013) found distinct patterns separating women who persist in engineering and those who opt-out using qualitative methods and categorized them as either individual or contextual factors. Factors like self-efficacy, self-confidence and other individual factors have been evidenced as reasons women do not persist (Bogue & Marra, 2009), but often at the expense of discounting the influence of external factors (Tharp, 2002). However, Fouad et al. (2016) later found no difference between the interests and workplace barriers experienced between women who stay in engineering and those who left. This potentially disruptive evidence highlights the importance of ensuring the appropriate supports are accessible to women at all levels. Table 5.2 captures a list of factors from the literature.

The Center for Gender in Organizations (1998) postulated four frames for understanding gender equity in organizations and was further explained by two of its original researchers. Table 5.3 shows all four frames and its benefits and limitations as described by Fletcher & Ely (2003).

Islam, Kafle, Wong, Guillaume-Joseph & Heart's (2017) study used a focus group of systems engineering professionals at various stages in their careers and having various degree attainment levels to develop a list of challenges for SE professionals and opportunities or conditions that were prevalent for SEs. These findings span four different themes, one being the need to focus on diversity. The findings associated with the fourth theme extracted into Table 5.4 echoes many findings of the research in the broader context of engineering. Systems engineers face very similar challenges as engineers at large but the nature and context of the systems work afford many opportunities that mitigate the risk and impact of those challenges.

The range of experiences among women has resulted in coping strategies commonly implemented by women who persist in engineering:

1. Find a "pocket of sanity" or a place where women can be themselves within the organization (Servon & Visser, 2011) and/or outside the organization.

TABLE 5.2

Support Factors and Barriers

Factors Impacting Women's Decision to Choose/Persist			
Factor	Source	Factor	Source
Workplace Climate and Experience	NAP (2006); Frehill (2008); Buse et al. (2009); Fouad & Singh (2011); Buse et al. (2013); National Academies Press (2014); Fouad et al. (2016)	Outcome Expectations	Fouad & Singh (2011); Singh et al. (2013); Fouad et al. (2016)
Travel Needs	Fouad & Singh (2011); Servon & Vissor (2011)	Recognition	Fouad & Singh (2011)
Family Obligations/ Work–Life Balance	Frehill (2008); Buse et al. (2009); Fouad & Singh (2011); Servon & Vissor (2011); Buse et al. (2013); Fouad et al. (2016)	Supportive People	Buse et al. (2009); Fouad & Singh (2011); Inda et al. (2013); Fouad, Singh, Cappaert, Chang & Wan (2016)
Interests	Fouad & Singh (2011); Inda et al. (2013); National Academies Press (2014); Fouad et al. (2016)	Role Models/ Examples	National Academies Press (2014)
Self-Confidence	Buse et al. (2009); Fouad & Singh (2011)	"Other" vs. "Self" Orientation	National Academies Press (2014); Buse et al. (2013)
Job Satisfaction	Fouad & Singh (2011); Singh et al. (2013); Fouad et al. (2016)	Organizational Commitment	Singh et al. (2013); Fouad et al. (2016)
Self-Efficacy	Buse et al. (2009); Buse et al. (2013); Inda et al. (2013); Inda et al. (2013); Fouad et al. (2016)	Turnover Intention	Singh et al. (2013); Fouad et al. (2016)
Engineering Identity/ Occupational Commitment	Buse et al. (2009); Buse et al. (2013); Fouad et al. (2016)	Bias in Performance Evaluation	Servon & Vissor (2011)
Adaptability	Buse et al. (2009); Buse et al. (2013)	Engagement/ Challenge	Frehill (2008); Buse et al. (2013); National Academies Press (2014)
Sexual Harassment	Servon & Vissor (2011)	Unwanted Attention due to Appearance	Servon & Vissor (2011)
Demeaning and Devaluing Exchange/ Being Viewed as Less Capable	Servon & Vissor (2011); Saavedra et al. (2014)	Authenticity/ Role Conflict	Fouad et al. (2016)

(Continued)

TABLE 5.2 (*Continued*)

Support Factors and Barriers

Factors Impacting Women's Decision to Choose/Persist			
Factor	**Source**	**Factor**	**Source**
Isolation	Servon & Vissor (2011)	Career Preferences	NAP (2006)
Competition	Servon & Vissor (2011)	Demographics	NAP (2006)
Goals/ Ambition	NAP (2006); Buse et al. (2009); Inda et al. (2013)	Financial Support	Inda et al. (2013)
Training and Development	Fouad & Singh (2011); Singh et al. (2013); Fouad et al. (2016)	Advancement Opportunities	Frehill (2008); Buse et al. (2009); National Academies Press (2014); Fouad et al. (2016)
Role Overload	Fouad et al. (2016)	Role Ambiguity	Fouad et al. (2016)
Rewarding Work	Buse et al. (2009)	Other Mentions	Exposure/Awareness, Fear/Unknowns, Individual Culture, Deductive Reasoning

TABLE 5.3

Four Frames on Gender

Frame 1—Fix the Women		Frame 3—Create Equal Opportunities	
Benefits	**Limitations**	**Benefits**	**Limitations**
Helps individual women succeed; creates role models when they succeed	Leaves system and male standards intact; blames women as source of problem	Helps with recruiting, retaining, advancing women; eases work–family stress	Has minimal impact on organizational culture; backlash; work–family remains "woman's problem"
Frame 2—Celebrate Differences		**Frame 4—Revise Work Culture**	
Legitimizes differences; "feminine" approach valued; tied to broader diversity initiatives	Reinforces stereotypes; leaves processes in place that produce differences	Exposes apparent neutrality of practices as oppressive; more likely to change organizational culture; continuous process of learning	Resistance to deep change; difficult to sustain

2. Disguise the differences between women and men (Singh et al., 2013); Behave like a man (Servon & Visser, 2011).

3. Make career moves based on comfort level rather than advancement potential, even if a lateral change or demotion is needed to obtain more tolerable work conditions (Servon & Visser, 2011).

4. Value feminity by highlighting and valuing psychological attributes that are considered feminine (Singh et al., 2013).

5. Present differences openly (Singh et al., 2013).

TABLE 5.4

Challenges and Opportunities Identified by Systems Engineers

Years of Experience in SE	SE Career Challenges	SE Career Opportunities
Fewer than 3 years	• Other engineering education streams at universities *lack SE perspective* and vice versa	• Ability to take more *leadership* role • Uniquely positioned to *mentor* others
3–9 years	• Limitation of *continuing education, skills, and techniques* that mature systems engineers over time • Lack of *leadership opportunity* for technical systems engineers	• Uniquely positioned to build *network and professional connections* that are helpful for successful career
10–19 years	• *Women engineers are less represented* in the overall science and engineering fields, including SE • Most *challenges* are not technical, but are *related to people, process, and culture* • SE is *less diverse* and is *male-dominated* engineering	• Opportunity for *growth* when systems engineers lead facilitation between engineering and management aspects of a project • Empowering future systems engineers through *SE education* has huge opportunity
More than 20 years	• Understanding the value of *continuous training and certifications*	• Opportunity to *lead* • Systems engineers *work more effectively with others*

6. Accept as normal; Navigate the existing norms and climate (Servon & Visser, 2011; Hahn & Lynn, 2017).

7. Consider trade-offs among alternatives to justify choice or the relational "best."

5.5 A Case Study

The following case study captures a mosaic of perspectives among women in ISE to broaden the narrative and individualize the topic. This output includes the identification of additional strategies used by the persisting women under study.

5.5.1 Methodology

A survey instrument was developed and used to ask 36 women in ISE about their decision to pursue engineering, persist and their outlook on success. Two men were included in the study because of their self-identification as eligible (total sample size = 38). Participants also took the general self-efficacy assessment (Schwarzer & Jerusalem, 1995). Professional experience ranged from those with no work experience in engineering (4%) to those with more than 30 years of experience (24%). The intermediate categories were $0 \le 5$ years (24%), $5 \le 10$ years (16%), $10 \le 20$ (12%), and $20 \le 30$ (20%). This experience spanned across industries including academia (56%), government (60%), non-profit (60%) and Private Organizations (64%), with many individuals having work experience in multiple industries.

Ninety-two percent of the participants identify as industrial or systems engineers, however the very nature of systems engineering can sometimes leave domain engineers wanting to identify within their core domain. The majority of participants are current professionals in engineering (84%), while 12% were students at the undergraduate and graduate levels and 4% were retired.

The survey was collected either online or in-person at ISE events to allow participants to contribute as much or as little as they do desire. Given the personal and sensitive nature of the input, no identifying information was collected, which limits the ability to analyze some aspects of the data gathered.

5.5.2 Results

The results are organized as three objectives: (i) identifying why participants chose engineering and (ii) continue to persist in engineering, and (iii) participants' outlook on having success in engineering.

5.5.2.1 Choosing Engineering

A large portion of participants made the decision to consider an engineering career in K-12 (44%), with K-8 accounting for about a third of these decisions (Figure 5.7).

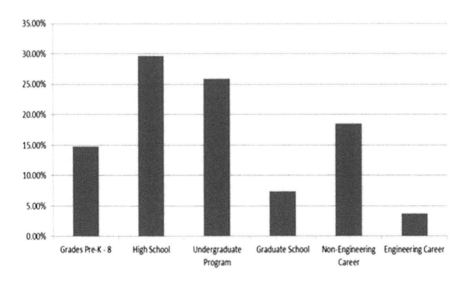

FIGURE 5.7
Point engineering became an option.

Table 5.5 captures the responses to the prompt to describe the most influential factors in the individual decision to become an engineer and shows a correlation to the literature. Women commonly cite role models, family and teacher encouragement/discouragement, an interest and propensity towards

TABLE 5.5

Choosing to Become an Engineer

Most Influential Factor(s) to Your Decision to Become an Engineer	
"I was involved in *several after-school programs* that introduced high school students to the technology and engineering"	"Science *teachers who encouraged me,* and a *general sense that I could do anything* I wanted to, imbued by *the way I was raised*"
"Ability to do technical drawing, designing, creating, and building, and to make decent money"	"My *father was an engineer* and he could support the family with that career. I knew *I could make an impact* in this field"
"INCOSE *organization*"	"*Family member encouragement*"
"I was involved with a *IT program in high school,* which exposed me to Web development, Networking and Robotics. I really *enjoyed the coursework, and seemed to do well* in the program"	"I always had an *interest in science and math.* I think the fact that *I was good at it,* influenced my decision to pursue engineering as a career"
"A *Co-worker* at a part time job I was working with at the bank, he was arising engineering technology and *made it sound fun*"	"The company I was currently *working for was an engineering company* and *supported (funded) my getting a degree in engineering*"
"*Job opportunities*"	"College degree availability and affordability"
"*Career options* and *high salary*"	"*a non-credit class*"
"I was in the School of Business, then took my first Math course under the Dean of Mathematics. I was her best student and *SHE told me that I should CONSIDER Math because I was very good* and it would also broaden my horizon! She was RIGHT!"	"I contacted the head of the Aerospace Engineering department at UCLA who was a woman and *asked her opinion on a future in engineering.* My *high school guidance counselor told me that it was a very bad idea and I had no positive future in engineering*"
"*Engineers, and especially female engineers I interacted with*"	"*Math and science abilities, Father's advice/ influence*"
"*Opportunity for Research Assistantship Interdisciplinary appeal* from technical, yet nonengineering background"	"*Dad was an engineer* I was *good at math and computer programming*"
"*Peers* and *experience*"	"*Taking a job* at MITRE"
"The *support of my family* and the *doubts of others*"	"*I liked and excelled at math and science* during my grammar and high school years"
"My dad offered to take me to work and I *met with women at his work about their choice* between Math (my current choice for college) and Engineering. Their responses convinced me to change to Engineering as a better fit to me as I was more interested in *hands on work* than theoretical work"	
"Relationship to my natural desires- I liked the idea of developing a toolkit that can be applied to many domains and especially liked the idea of "making things better". Awareness- Once I knew what industrial engineering was, the *opportunities inherent* to that become an engineer, my decision was made. However, if the lady in the office had never field and the requirements to approached me to *suggest that I challenge myself,* I would have continued on my journey to be a lawyer and never discovered this option"	

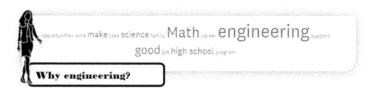

FIGURE 5.8
Keywords from prompted response choice.

FIGURE 5.9
Keywords from prompted response moment.

mathematics, science and hands-on applications, and exposure to the field as heavily influencing their decision to choose engineering.

Other intrinsic factors (i.e., self-confidence, perceived humanitarian impact) and extrinsic factors (i.e., salary, opportunity, funding) are captured in the participants' response. There are several instances where the participants were on an alternative path and external influences caused a pivot to the field of engineering. Yet, the main factors were as follows: examples (other women/parents (dad)); self-confidence; prospect of impact; others seeing potential; doubt; peers; professional organizations; accessibility (degree availability/affordability); and workplace. The word cloud shown in Figure 5.8 captures the most cited words and phrases in participant responses:

In addition, participants' described the moment they decided to become an engineer (Table 5.5). The main factors were as follows: examples (other women/parents (dad)); self-confidence; interests; exposure/awareness; propensity to math/science; coursework; funding/development opportunities; opportunity; good teachers; authenticity; familiarity of people; relational; encouragement; experiential opportunities (internship); deduction; organization; and inspired as inferred from Table 5.5. The word cloud in Figure 5.9 captures the most cited words and phrases in participant responses (Table 5.6).

5.5.2.2 Persistence in Engineering

Many researchers have found evidence of the impact of self-efficacy in the decision to persist. A validated instrument used to measure the general self-efficacy of an individual showed relatively consistent, aggregate levels for each measure. Participants felt that each item was between "Moderately true" and "Exactly true," and proved consistent across each group of participants (i.e., students, professionals, and retirees).

TABLE 5.6

Moment of Choosing Engineering

Moment You Made the Decision to Become an Engineer	
"It was after *talking with several engineers* about the *interesting projects* they worked on in their career."	"Freshman year in undergrad; wanted to start career directly after graduation."
"When I was accepted to MIT. I figured if they thought *I had a shot* at it, that I probably could."	"After a summer assistantship, I was awarded a graduate assistantship and *offer for scholarship in engineering.*"
"On campus, not knowing what to major in to truly satisfy my *personal desires*. My girl friend from high school walks by, and *invited me to go to a lecture* on engineering, I did, when I got there, I knew and heard the conversation on *possibilities*, I was knew I was home!"	"The NASA moon landing *video*. At that moment I realized that the Earth was not the only thing in the solar system and that there was a thing known as a solar system. As a young child, I wanted to build my own spaceship and live on another planet."
"7th grade, when I had an *amazing math teacher* that taught in a way I understood. Before I was a C student, and overnight after taking his class, I was ranked #1. Engineering then seemed possible."	"I applied for both investment banking jobs and engineering jobs when I graduated from college. *Received engineering offers—hence became an engineer.*"
"When I understood the thought processes behind systems and *how effectively I could and do apply* these processes/steps to solve problems."	"Looking through the course catalog when I was *visiting colleges and reading the course descriptions* of the Optics courses."
"I chose my school based on the fact that it had engineering options, at the end of high school I was not considering colleges without Engineering programs. So there was *no specific moment so much as an accumulation of decisions.*"	"I decided to become an engineer when I took my first *internship* at UCLA Institute of Interplanetary Physics as an Aerospace Engineer and the head of the department told me that he wanted to *mentor* me throughout my undergraduate career and help me pursue my dreams of working on a Mars project."
"When I met *friends* in the engineering program *with similar background* which did not prepare us for any STEM career field, *I knew I could do it too.*"	"In college, *I was told* that I could either teach or become an engineer with a Mathematics degree. *I didn't have the patience to be a teacher.*"
"I was always *good at math and science*. The *interview with other women* was what really changed my mind. Before that I just knew I was really, really good at math so I was going to major in math."	"For about 18 months, I had a *dual career* in the government as a Mathematical Statistician and a Computer Science Programmer. That spear-headed my career into engineering and IT."
"When I enrolled in Computer Science *classes* in college."	"High school"
"Family member brought me a *SWE* magazine."	"*Joining a* Systems Engineering *organization*"

(Continued)

TABLE 5.6 (*Continued*)

Moment of Choosing Engineering

Moment You Made the Decision to Become an Engineer	
"On my third day of work with the company [I worked for]. That was the earliest day I could speak with someone in authority who could answer my question on how quickly I could use the company *educational benefits.*"	"Several of my *colleagues are engineers* and one was pursuing her PhD when we had *a conversation that planted the seed* in my mind that I, too, should think about *pursuing more in terms of my education and career.*"
"I wanted to have a *marketable skill* after military retirement."	"Working on my masters degree"

"*Someone expressed* to me during a College Work Study program *that I should challenge myself* instead of pursuing a degree in political science. My K-12 experience in an impoverished community had only exposed me to the options of doctors, nurses, lawyers, teachers and athletes so I had a very *limited understanding of the opportunities.* Fortunately, this comment made me step back and think about the prospects of pursuing something new. I *reviewed the list of engineering majors* offered by my University and selected *a major that conformed* to me most- industrial engineering. I liked the idea of developing a toolkit that can be applied to many domains and especially liked the idea of 'making things better'."

Participants then weighed in on questions that aim to understand their persistence in engineering. Figure 5.10 shows that among those factors measured, participants feel that self-confidence and the influence of peers and colleagues are the most influential factors in the belief that they can be successful in engineering.

Table 5.7 captures the prompted response of each participant and further substantiates these findings. Similar to the decision to pursue engineering, both intrinsic and extrinsic factors are present in the form of both supports

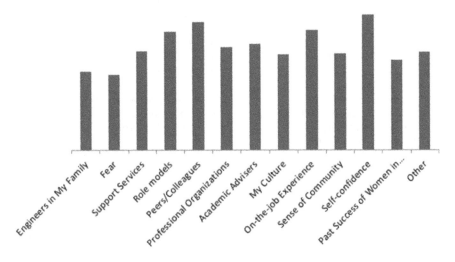

FIGURE 5.10
Factors influencing the belief of successful outcome.

TABLE 5.7

Factors Influencing the Belief of Successful Outcome in Engineering

Most Influential Factor(s) to the Belief That You Can Be Successful in Engineering	
"Advisors and role models"	"*Role models* in academia and in the industry"
"*Dad was an engineer* and *told me* I could be a successful engineer; Now I *see women with families* who are currently successful engineers"	"I *succeeded easily* in classes in high school most similar to engineering. I was also *told by parents and advisors* in high school that *it would be a good career path.*"
"Just going for it without knowing how much math was involved, or complex sciences, I viewed only the *potential outcome*"	"*Exposure* to Engineering, *Confidence* (math/science), *Passion* for problem-solving and innovative concepts, *funding*"
"My *performance*; My company has *promoted me continuously and provided encouragement*"	"How *others interact with me.* If I feel others respect and value my work, I feel I am succeeding."
"Getting my masters in SE"	"MITRE"
"Just pure *ignorance that I might not be successful.* I had always earned high marks in school and it was not difficult for me, I was able to spend time in sports and doing things I liked which included homework."	"The most influential factor that I can be successful is the *women I see in my workplace*, that are doing phenomenal things in the field of engineering. Also, *my cohort* and the cohorts before me are awesome in their own right and they have been an *inspiration and help to me!*"
"First *internship*, actually doing work that is considered engineering work"	"*Self-confidence* and *excelling in math, algebra, geometry, trigonometry*"
"Having the *support of your peers* (NOT working in an environment where you skills are constantly questioned)"	"Small successes doing what other male counterparts were known for"
"My perception that there are problems facing our country (and the world) that need solving, to which I *think I can contribute.*"	"My belief that I could be successful was really rooted in the fact that all of the *external research* I did on my own on topics that I *was interested in was co-written by women.*"
"When someone *tells me I can't, I do,* especially when it is a man who tells me I can't."	"The *support and encouragement of family, friends and colleagues.*"
"The *support of my family* and the determination to tell my kids that they can become anything they want to become, then *stand as that example.*"	"*Work harder and smarter than the male engineer* you sit next to; *use your female traits* like intuition and soft skills to *be a better engineer*"
"*Confidence* and *preparation*"	"*On the job experience*"
"*Self-confidence, Mentors, Experience*"	

"I think that it is the internal belief that *I can do anything.* I live on the premise that I can only do the best that I can and that whatever outcome (success or failure) that prevails will forge a better me. In this school of thought there will always remain an outlook of success as long as I persist. I do acknowledge that the journey ahead will require a tremendous amount of strength, as many days are a strain but given the plethora of experiences in my life, I have no doubt that *success is imminent.*"

FIGURE 5.11
Keywords from prompted response influence.

and barriers. The main factors were confidence, mentors, experience, recognition, examples (other women and parents [dad]); outcome expectations, propensity for math/science; peer interaction; relativity to men; experience; valuing femininity; and experiential learning (e.g., internships). The word cloud shown in Figure 5.11 captures the most cited words and phrases in participant responses:

Table 5.8 extends this inquiry by asking participants about specific challenges to the belief that they can be successful. The word cloud shown in Figure 5.12 captures the most cited words and phrases in participant responses:

TABLE 5.8

Challenges to the Belief of a Successful Outcome in Engineering

Challenges to the Belief That You Can Be Successful in Engineering	
"Weeder classes in college: I failed Intro to EE twice due to *awful professors* (they intentionally failed 90%+ of each class). I *almost switched my major* to business. But after the 1st two years, I *emerged as a technical leader* amongst my peers."	*"Time* is always a challenge in that it is a very precious, limited commodity. I am *wife, mother,* and involved with *other extracurricular activities* that I love and want to be a part of. So *managing and prioritizing my time* is critical to achieving my goals."
"That *maybe you weren't that good,* or *not interested in being that good* in math and or science."	*"Any setback,* especially at work (hasn't happened yet, but if it did, it would challenge that belief)."
"Being the only female in undergraduate courses was tough. Particularly *if I did not do well* on an assignment, *it made me feel isolated and discouraged."*	"I was *fairly confident* in my ability to succeed as an engineer. *Some classes/professors* that I struggled with in college *made me doubt that but not enough to quit."*
"I have been *in a very challenging position* and *many people in my organization were disrespectful of me.* I was *made to feel I was failing* and that *I was the problem.* The reality was not that at all, (I realize now) but at the time that environment was *very destructive for me.* I actually ended up *clinically depressed* for about 2 years."	"One challenge to this belief was the belief of some *stakeholders* I encountered who *doubted that a young female could really help* much. This challenge did not discourage me too much because *I had bosses and colleagues willing to let me try."*

(Continued)

TABLE 5.8 (*Continued*)

Challenges to the Belief of a Successful Outcome in Engineering

Challenges to the Belief That You Can Be Successful in Engineering	
"My *peers and my professors*. I have had more than one *professor* that *questioned my desire to be an engineer*. Additionally, I have had *classmates* that *avoided working with me and tried to literally set me up for failure*."	"*I took nine years off work to raise children*. There were several years that *I did not believe I would be able to enter back into the workforce as an engineer*. I thought I would lose my technical skill and that I wouldn't be hired."
"*Lack of experience*: As a recent graduate, I'm undecided if I want to be in academia or industry."	"*Bias from others*. Some people believe *women and other minorities are inferior and incompetent*."
"Every now and then I look up and look around. There is often *no one who looks like me from the respect of race, gender and other common measures*. I find that a great deal of *conformity must occur* from the way that I speak, to the way that I choose to do my curly natural hair, and to the way that *I feel I am received*. I often feel that *I have to first prove that I am qualified before I am even acknowledged as being credentialed*, despite having a terminal degree, years of experience and several state license and certifications in the field. This can be *exhaustive and poses a threat* to my overall desire to persist."	"Apparently in looking back *others* tried to *put up obstacles* (I'm going to give you an 'F' in mechanical drawing in high school - I was the only girl in the class). *Feedback from others* that I need to be *concerned about my image* (didn't know what they were talking about). *Competition from other women* that I didn't even know I was competing with (because I wasn't). Mostly I was *oblivious* to these things and just really enjoying life."
"*Lack of promotion and/or leadership opportunities*"	"I was *not accepted to my college in engineering*, but was in liberal arts"
"Peers feeling the need to *step on others in order to shine*."	"Not knowing as a woman *if I would be allowed or accepted*"
"*Communities and populations that lack exposure to engineering and careers in engineering* may not even consider engineering, much less imagine being successful in the field. *Funding/support* can also be a great challenge, as many who may aspire to succeed in engineering may be *overwhelmed with other responsibilities* and *less able to devote resources to the discipline*."	"I tried to pursue creating a publication while in undergraduate studies and I *could not find any engineers willing to work with me* and take a chance on having a woman's name on the paper. For example, I asked my statics and dynamics professor to assist me with writing a workbook of Aerospace specific examples and *he said that women did not the experience in that field* to pursue such an endeavor."
"Certain *cultures are not supportive of women* in general. Unfortunately, men from these cultures are in engineering. Therefore, it is very challenging to be successful in engineering if you are being taught by, or must work with, men from these *suppressive cultures*."	"Definitely *discrimination* in the field of engineering; I have been doing this for the past 33 years so I have seen a tremendous change in how female engineers are treated and respected but we still *don't garner the respect an equivalent male engineer (years experience etc) gets*"
"*Bias* about *not having an engineering degree*"	"*Lack of a support community*"

FIGURE 5.12
Keywords from prompted response challenges.

5.5.2.3 Outlook in Engineering

The final area of the case study examined participants' outlook on success in engineering. Despite some participants mentioning the fear of starting over, or other proxies of fear in their written responses, "fear" was the least influential factor among those explicitly queried. Career outlook, self-confidence, and on-the-job experiences were rated highest among participants.

About 80% of participants expressed that they had experienced a negative sense of belonging in engineering at some point and, in cases where they have overcome that state, shared their strategies for overcoming as shown in Tables 5.9 and 5.10. Others expressed both optimistic and pessimistic views of their outlook on belonging in their response.

The word clouds shown in Figures 5.13 and 5.14 capture the most cited words and phrases in participant responses to factors causing persistence and coping strategies, respectively.

The case study presents unfiltered, prompted responses for all questions ranging from 2 to 219 word responses per individual that show the intricate differences among each person's perspective.

The result yields validity to the findings of the literature in that the experiences of women in engineering is a complex and confounding interplay

TABLE 5.9

Factors Causing Persistence

Most Influential Factor(s) Contributing to Your Sustained Pursuit of the Engineering Field	
"The idea of *solving technical problems*"	"*Salary* and *career flexibility*"
"*Feeling supported by peers.* INCOSE has been a lifesaver for me when my work environment was very negative."	"The opportunities for *intellectual challenge* and *career growth* that are presented by doing so."
"The *determination to succeed* and *be accomplished*"	"Numerous *career option*"
"*Recognition; Hard work*"	"*Motivation from colleagues* around me."
"The *diversity and challenge of the work* in engineering. There is never a "mundane" day—*always learning and challenging*."	"*Financially support my family* doing *work I enjoy. I can contribute* and I can *keep learning*."

(Continued)

TABLE 5.9 (*Continued*)

Factors Causing Persistence

Most Influential Factor(s) Contributing to Your Sustained Pursuit of the Engineering Field	
"Engineering salary"	*"Career opportunities"*
"Knowing that I can succeed. Challenging myself to *defy all odds* and the limited expectations set by others in order to *extend my current peak*. I am quite young and through this journey through engineering, I better understand myself. I better understand other people. I have experienced ups and downs that one by one are shaping my final outcome. Not sustaining my pursuit means that I never uncover what/who I could have been; Another influential factor is that I have conversations with people who sometimes express the *importance of my persistence in their decision to pursue or persist.* I find that many women and girls tend to perceive me as inspirational, genuine and very approachable. Unlike what they may have experienced with others in the field, I listen to them beyond the technical matter, find aspects to relate and tend to leave an impression that has a positive impact on their academic and professional choices."	*"Family encouragement;* Growing up *surrounded by people who were very mechanical who had a wide range of knowledge about so many things; Curiosity,* there is so much you can self-study in other fields or subjects, but understanding engineering and how the world works takes education. There is *always more to learn and understand* in this field."
"Salary and *work/life balance"*	*"Family obligations;* the amount of *time spent in this field; fear of starting over"*
"The overall *drive to have an impact* in helping people with their problems and the desire *to do impactful meaningful work* keeps me going when I decide to do things outside of my comfort zone which mostly have to do with relating to other people a lot. I much like working alone or alongside others in a cooperative manner. I get *very discouraged when others are competitive with me."*	"Once I began my Aerospace specific curriculum in my undergraduate studies the number of women in the program dropped down from over 20 to 4. My need to meet the challenge to *finish what I started* plus the *immense support from my co-workers* in the physics lab pushed me to pursue engineering."
"My *self-confidence* and *understanding that my peers are no more prepared or knowledgeable than I am."*	*"Continued growth and success* in applying and continued learning of SE, *alignment with professional organizations"*
"Career outlook, work satisfaction"	*"Being competitive* and *love the challenge"*
"I'm *driven to succeed* in whatever I start. I was not classically trained as an engineer. Entering the field later in my career is difficult as people don't really consider you an engineer, so it is *hard for me to call myself one.* It's a shame."	"My *employer, fellow employees/peers, family, friends, cohort, professors are all positive influential factors* in my pursuit of my terminal degree and career in engineering."
"Funding and *family support.* Also a *passion for the subject matter."*	"I like *solving complex problems;* I like thinking logically and systematically."

TABLE 5.10

Coping Strategies by Participants

Strategies Used to Overcome the Negative Sense of Belonging You Have Felt	
"Dumb luck: I almost left the field to go get an MBA about 5 years into my career. I was not being given good work so I assumed engineering was just not for me. Then *some good work turned up and I was able to show my skills and work ethic* and everything changed. At other times when I was not being treated respectfully *I just found other venues (like INCOSE) where I could feel my knowledge and skills were valued*."	"I try to balance my personal and professional self, I feel that I am often forced to neglect one part of me in order to enjoy the other. Given the differences among the two, I tend to go through a cyclic sense of belonging and am really unsure if it will ever truly balance."
"College weeder *classes*"	"*Encouraging words from mentors*"
"*Inner Self-Confidence-* If I make it my problem, then I fail myself; Knowing that for some situations, *I won't participate and that's OK*."	"*Find organizations and colleagues, both cultural and gender specific*, to provide a positive sense of belonging."
"I realized that what I once perceived as lack of foundation or unorthodox thinking was truly beneficial in my *quest for knowledge* and also my way of solving problems."	"Most of the time it has been other people putting this doubt on me, so I just try to make sure that I *remember I was able to get my degree and job based on my own merit* so I belong where I am."
"$$, need to support family"	"*Discussions with mentors and peers*"
"I haven't. But I often *compare myself to peers - if they can do it, I can too*."	"Difficulty in being a recognized DE resource led me to *complete a MS in SE to gain confidence and recognition*"
"*I left unsupportive organizations for increasingly better organizational cultures with greater opportunities*, where I felt my *skills were appreciated* and I felt that *I was delivering high quality work*."	"I am still working on overcoming that. I have been a victim of intense abuse and harassment and it will take a while to move past that. With the *help of very concerned and family-like mentors*, I am working on this everyday."
"*Buried the negative, focused on the positive*"	"I remind myself that *I am where God means me to be*."
"I just tried to *focus on my strong skills*, and reminded myself only I could bring that unique combination of skills to the table."	"Just kept *trying to be a better engineer than my male counterparts*; made it a *personal challenge to get the best review, the highest raise*, etc."
"*Continued success*"	"By *knowing that I CAN do the work* and that I *AM smart*."

(Continued)

TABLE 5.10 (Continued)

Coping Strategies by Participants

Strategies Used to Overcome the Negative Sense of Belonging You Have Felt	
"I've had to *turn to people to get their perspective,* and to *block out my own negative thinking or memories of men telling me I didn't belong in the field* (or the classroom)."	"*Partake in outreach working with youth,* usually STEM related. *Investigate pay time business opportunities* and *doctoral programs to teach* instead."

"I never did overcome a negative sense of belonging, I don't really belong anywhere. In the sense that I am a woman who is introverted, analytical, like working on technical and what others see as mundane analyses, and I feel very alone - I don't really know anyone like me. Other women seem to want to compete and men just don't want to take me seriously or want me to behave a certain way that they expect women to behave where I've spent most of my life working and socializing with men. It's a shame I've come to this conclusion after decades in the field - I didn't start out this way - in the beginning *I was impactful, learning, making a difference, collaborating,* but others who were not as technically competent or as competent from a business sense would get moved up and I would help them (because they needed it) but more and more others have taken credit for my work or want me to do their job for them. I have my own work and dreams and pursuits, but I do want to *collaborate with those who are willing to be in it together and are willing to put in the time and do the work with me and are not just looking for the credit.*"

FIGURE 5.13
Keywords from prompted response motivation.

FIGURE 5.14
Keywords from prompted response strategies.

among internal, external, and greater societal factors. Women expressed views and experiences along their journey of persistence that are compounded by a relational and malleable network of factors and constructs. No individuals reported the same experience. However, themes were present across both efforts.

The factors compiled from the literature review (see Table 5.2) were used to gauge the level of traceability among the sample participants. The list is not all encompassing but does include the majority of common factors present in the research reviewed. All factors reported from the literature herein were found in the ISE case study.

5.6 Professional Support

Studies in the literature acknowledge the variety of factors women face in engineering and the relationships among those factors. Of the many funding opportunities to further study this topic and develop interventions that have theoretical grounding, empirical evidence, and merit and broader impacts, fewer studies provide resources for women at their individual and immediate locus. What can women do in the near term to improve their personal and professional development, outlook, and likelihood of persisting in engineering? What resources are available to empower women to be successful in the engineering space? How can the power of a critical mass around the topic result in a paradigm shift in organizational culture and accelerate the journey to gender parity in engineering?

Engineering's multi-faceted landscape requires inputs from technical and social perspectives in order to optimize both the physical and nonphysical aspects of engineered systems. Optimal utilization of human capital can result in better solutions to complex problems and the overall health and well-being of a more cooperative engineering labor force. The ways women position themselves in this professional culture and cope with the gendered discourse they experience have a significant impact on their decision to persist (Saavedra et al., 2014).

Kasperovich (2016) claims that "a persistent woman can overcome both the perceived and actual difficulties [of engineering] and gain fulfilling experiences in any engineering profession." Presumably, a "one size fit all" solution as to how this occurs is infeasible. Rather than assuming this homogeny as a precondition, the experience, views, and strategies of many others are provided. It is impractical to suggest which are better than others without consideration of each individual's internal and external factors; therefore, no attempt to do so is present in this work.

However, professional memberships provide an opportunity to network, learn new things, gain leadership experience, and ultimately stay adept in some area. The case study showed that a little over 50% of the participants participate in some sort of women-centric engineering opportunity or professional organization, but more than 30% would like to increase their existing participation level. Failure to do so is often a lack of time to artfully inquire or sift through results to identify the "right" information or opportunity for the individual and their specific circumstances.

Table 5.11 provides a list of women-centric organizations and initiatives that serve either the ISE community specifically or the broader

TABLE 5.11

Women-Centric Organizations for ISE Community

Organization/Effort	Website
ASEE Diversity, Equity & Inclusion Committee	http://diversity.asee.org
Association for Women in Science	www.awis.org
Campaign for Gender Balance in Science, Technology, and Engineering, UK	www.wisecampaign.org.uk
Engineer Girl	www.engineergirl.org
IEEE Women in Engineering	http://wie.ieee.org
IEEE Women in Engineering, United Kingdom and Ireland	www.ieee-ukandireland.org/affinity-groups/women-in-engineering/
IISE Minority and Women Industrial Engineer Outreach Program	www.iise.org/details.aspx?id=46020
INCOSE Empowering Women as Leaders in System Engineering (EWLSE)	www.incose.org/ChaptersGroups/initiatives/ewlse
International Network of Women Engineers and Scientists (INWES)	www.inwes.org
NDIA Women in Defense (WID)	www.womenindefense.net
NSF INCLUDES: Inclusion across the Nation of Communities of Learners of Underrepresented Discoverers in Engineering and Science	www.nsf.gov/news/special_reports/nsfincludes/index.jsp
NSF Increasing the Participation and Advancement of Women in Academic Science and Engineering Careers (ADVANCE)	www.nsf.gov/funding/pgm_summ.jsp?pims_id=5383
Sally Ride Science @ UC San Diego	https://sallyridescience.ucsd.edu
Society of Women Engineers	http://societyofwomenengineers.swe.org
The 50k Coalition: 50,000 Diverse Engineers Graduating Annually by 2025	http://50kcoalition.org
The European Association for Women in Science, Technology, Engineering and Mathematics	www.witeceu.com
Women and Girls in STEM	https://energy.gov/diversity/services/women-and-girls-stem
Women in Information Technology Science and Engineering	www.it-ology.org/get-connected/programs/women-in-it/
Women's Engineering Society	www.wes.org.uk
Women@NASA	https://women.nasa.gov
American Association of University Women	www.aauw.org
Women in Science, Engineering and Medicine of National Academies	http://sites.nationalacademies.org/pga/cwsem/
Women in Science and Engineering (WISE)	www.nsbe.org/Professionals/Programs/Special-Interest-Groups-(SIGs)/Women-in-Science-Engineering-(WiSE).aspx#.XHg9IpxKhaQ

engineering community. Each organization offers a range of opportunities that may include mentor programs, webinars, training, conferences, networking, publications, and other resources.

For example, Empowering Women as Leaders in Systems Engineering (EWLSE) is a women-centric group under the INCOSE. It meets during the INCOSE International Workshop and INCOSE International Symposium each year to provide an opportunity for women to build relationships; discuss issues women face in engineering; and devise strategies to reach more participants and increase the likelihood of individual, organizational, and global impact. This group is one of many that have made deliberate efforts to build relationships among women in engineering and purposefully address the gender gap.

Many of the organizations offer more than one "opportunity," and voluntary participation tends to be at the level an individual deems appropriate for their circumstance. Furthermore, many of the organizations have an extensive, historical presence in the field and have reoccurring opportunities while others have a more fixed, short-term objective.

Table 5.12 captures the products of efforts on this topic to provide a range of resources that may help to understand and combat some of the factors women face in engineering. Additional resources are provided in the appendix (Table 5.13).

5.7 Conclusion

The findings of this work may challenge the thinking of many readers because it aims to acknowledge the challenges and opportunities that many women face in their journey to persist in engineering. It is also an attempt to acknowledge both progress in the field and fallacies in blindly neutralizing the systemic discourse embedded in the profession. It is only when steps are taken to recognize that discourse exists, that engineers can employ effective strategies and seek the resources needed to foster their persistence in a system that will take some time to truly change.

The best conjecture for how this occurs is by starting with this work and the work that precedes to gain a more holistic and informed lens. This approach may be met with some resistance across the board, but it stimulates increased awareness of the conditions of the ISE profession and engineering in general and highlights available resources. The intended message is that irrespective of the factors forging each individual's current state in engineering, there is a web of support structures strengthening from this discourse, and through each woman's persistence, gender parity in engineering is inevitable.

TABLE 5.12

Resources for Women in ISE

Resource	Website
Women in Industrial Engineering; Stereotypes, Persistence, and Perspectives	http://onlinelibrary.wiley.com/doi/10.1002/j.2168-9830.2012.tb00051.x/abstract
The Ten Steps for Sustaining the Pipeline of Female Talent in Science, Technology, Engineering and Manufacturing (STEM)	www.wisecampaign.org.uk/consultancy/industry-led-ten-steps/10-steps
Resources for Women in the STEM Fields: Books and Online Sources	https://grccwomenstudies.files.wordpress.com/2010/09/women-in-stem-fields-resources.pdf
Impact $10 \times 10 \times 10$ Gender Parity Report	http://online.fliphtml5.com/zmam/ndms/#p=156
WEPAN Framework for Promoting Gender Equity in Organizations	www.wepan.org/?page=FourFrames
Women Matter 2016: Reinventing the Workplace to Unlock the Potential of Gender Diversity	www.mckinsey.com/~/media/mckinsey/global%20themes/women%20themes/women%20matter/reinventing%20the%20workplace%20for%20greater%20gender%20diversity/women-matter-2016-reinventing-the-workplace-to-unlock-the-potential-of-gender-diversity.ashx
Time for a New Gender-Equality Playbook	www.mckinsey.com/global-themes/leadership/time-for-a-new-gender-equality-playbook?
Diversity Matters Report	www.mckinsey.com/~/media/mckinsey/business%20functions/organization/our%20insights/why%20diversity%20matters/diversity%20matters.ashx
Why Diversity Matters Article	www.mckinsey.com/business-functions/organization/our-insights/why-diversity-matters
Delivering through Diversity Report	www.mckinsey.com/~/media/McKinsey/Business%20Functions/Organization/Our%20Insights/Delivering%20through%20diversity/Delivering-through-diversity_full-report.ashx
Delivering through Diversity Article	www.mckinsey.com/business-functions/organization/our-insights/delivering-through-diversity
Women in the Workplace 2017 Report	https://womenintheworkplace.com/Women_in_the_Workplace_2017.pdf
Women in the Workplace 2017 Article	https://womenintheworkplace.com

(Continued)

TABLE 5.12 (*Continued*)

Resources for Women in ISE

Resource	Website
Reinventing the Workplace for Greater Gender Diversity	www.mckinsey.com/global-themes/gender-equality/reinventing-the-workplace-for-greater-gender-diversity
Women in the Workplace 2016 Article	www.mckinsey.com/business-functions/organization/our-insights/women-in-the-workplace-2016
How Companies Can Guard against Gender Fatigue	www.mckinsey.com/business-functions/organization/our-insights/how-companies-can-guard-against-gender-fatigue
Fostering Women Leaders: A Fitness Test for Your Top Team	www.mckinsey.com/business-functions/organization/our-insights/fostering-women-leaders-a-fitness-test-for-your-top-team
Addressing Unconscious Bias	www.mckinsey.com/business-functions/organization/our-insights/addressing-unconscious-bias
Championing Gender Equality in Australia	www.mckinsey.com/business-functions/organization/our-insights/championing-gender-equality-in-australia

Appendix: Additional Professional Support for Women in Engineering

TABLE 5.13

Professional Support for Women in Engineering

Additional Resources/Reports

"Athena Rising: How and Why Men Should Mentor Women" By W. Brad Johnson, David Smith, © 2017—Routledge.

Ambrose, Susan A. Journeys of Women in Science and Engineering: No Universal Constants. Philadelphia PA: Temple University Press, 1997.

Bystydzienski, Jill M., and Sharon R. Bird. Removing Barriers: Women in Academic Science, Technology, Engineering, and Mathematics. Bloomington: Indiana University Press, 2006.

Committee on Maximizing the Potential of Women in Academic Science and Engineering, and Committee on Science, Engineering and Public Policy. Beyond Bias and Barriers: Fulfilling the Potential of Women in Academic Science and Engineering. Washington DC: National Academies Press, 2006.

Committee on Maximizing the Potential of Women in Academic Science and Engineering, et al. Biological, Social, and Organizational Components of Success for Women in Academic Science and Engineering: Report of a Workshop. Washington DC: National Academies Press, 2006.

Davis, Cinda-Sue. The Equity Equation: Fostering the Advancement of Women in the Sciences, Mathematics, and Engineering. 1st ed. San Francisco CA: Jossey-Bass Publishers, 1996.

Long, J. Scott, and National Research Council, Committee on Women in Science and Engineering. From Scarcity to Visibility: Gender Differences in the Careers of Doctoral Scientists and Engineers. Washington DC: National Academy Press, 2001.

National Academy of Engineering, Committee on Diversity in the Engineering Workforce. Diversity in Engineering: Managing the Workforce of the Future. Washington DC: National Academy Press, 2002.

National Research Council, Committee on Women in Science and Engineering, Office of Scientific and Engineering Personnel. Who Will do the Science of the Future?: A Symposium on Careers of Women in Science. Washington DC: National Academy Press, 2000.

National Research Council, Committee on Women in Science and Engineering, and National Research Council, Committee on Policy & Global Affairs. Female Engineering Faculty in the US Institutions: A Data Profile. Washington DC: National Academies Press, 1998.

Teaching the Majority: Breaking the Gender Barrier in Science, Mathematics, and Engineering. New York: Teachers College Press, 1995.

Schiebinger, Londa. Gendered Innovations in Science and Engineering. Stanford CA: Stanford University Press, 2008.

Shaywitz, Sally, Jong-on Hahm, and National Research Council, Committee on Women in Science and Engineering. Achieving XXcellence in Science: Role of Professional Societies in Advancing Women in Science: Proceedings of a Workshop AXXS 2000. Washington DC: National Academies Press, 2004.

Stewart, Abigail J., Janet E. Malley, and Danielle LaVaque-Manty. Transforming Science and Engineering: Advancing Academic Women. Ann Arbor MI: University of Michigan Press, 2007.

Williams, F. Mary, and Carolyn J. Emerson. Becoming Leaders: A Practical Handbook for Women in Engineering, Science, and Technology. Reston VA: American Society of Civil Engineers, 2008.

References

ABET (2018). Accreditation Board for Engineering and Technology. Retrieved from www.abet.org/.

Adcock, R., Squires, A., Gannon, T. & Virani, S. (2015). Systems engineering education for all Engineers. *2015 IEEE International Symposium on Systems Engineering (ISSE)*, Rome, 2015, pp. 501–508.

Atman, C., Sheppard, S., Fleming, L., Miller, R., Smith, K., Stevens, R. & Lund, D. (2008). Moving from pipeline thinking to understanding pathways: Findings from the academic pathways study of engineering undergraduates. *2008 ASEE Annual Conference*, Pittsburgh, PA, June, 2008.

Bogue, B. & Marra, R. (2009). Help her believe in herself. *The ASEE Prism*, 18(8), 49.

Brawner, C. E., Camacho, M. M., Lord, S. M., Long, R. A. & Ohland, M. W. (2012). Women in industrial engineering: Stereotypes, persistence, and perspectives. *Journal of Engineering Education*, 101, 288–318.

Bureau of Labor Statistics (2016). Employment and Wages of Engineers in 2015. Retrieved from www.bls.gov/opub/ted/2016/employment-and-wages-of-engineers-in-2015.htm.

Buse, K., Billimoria, D. & Perelli, S. (2013). Why they stay: Women persisting in US engineering careers. *Career Development International*, 18(2), 139–154.

Buse, K., Perelli, S. & Bilimoria, D. (2009). *Why They Stay: The Ideal Selves of Persistent Women Engineers*. Cleveland, OH: Case Western University.

Cadaret, M., Hartung, P., Subich, L. & Weigold, I. (2017). Stereotype threats as a barrier to women entering engineering careers. *Journal of Vocational Behavior*, 99, 40–51.

Cech, E., Rubineau, B., Silbey, S. & Seron, C. (2011). Professional role confidence and gendered persistence in engineering. *American Sociological Review*, 76(5), 641–664.

Center for Gender in Organizations (1998). *Making change: A Framework for Promoting Gender Equity in Organizations. CGO Insights*. Boston, MA: Simmons Graduate School of Management.

Duncan, J. & Zeng, Y. (2005). *Women: Support Factors and Persistence in Engineering*. National Center for Engineering and Technology Education. http://ncete.org/flash/research/Report%20_Yong-Duncan_.pdf.

Fletcher, J. & Ely, R. (2003). Introducing gender: Overview. In R. Ely, E. Foldy, M. Scully, and the Center for Gender in Organizations, Simmons School of Management (Eds.), *Reader in Gender, Work, and Organization* (pp. 3–10). Hoboken, NJ: Wiley.

Fouad, N. & Singh, R. (2011). Stemming the Tide: Why Women Leave Engineering. Retrieved from https://energy.gov/sites/prod/files/NSF_Stemming%20the%20Tide%20Why%20Women%20Leave%20Engineering.pdf

Fouad, N., Singh, R., Cappaert, K., Chang, W. & Wan, M. (2016). Comparison of women engineers who persist in or depart from engineering. *Journal of Vocational Behavior*, 92, 79–93.

Foucault, M. (1995). *Discipline and Punishment* (reprint). New York: Vintage Books.

Frehill, L. (2008). Why do women leave the engineering workforce? *SWE Magazine of the Society of Women Engineers*, Winter, 24–26.

Garbee, E. (2017). The Problem with the Pipeline: A Pervasive Metaphor in STEM Education has Some Serious Flaws. Retrieved from www.slate.com/articles/technology/future_tense/2017/10/the_problem_with_the_pipeline_metaphor_in_stem_education.html.

Government Accountability Office-GAO (2016). Strategies to Address Representation of Women Include Federal Disclosure Requirements. Retrieved from www.gao.gov/assets/680/674008.pdf.

Government Accountability Office-GAO (2018). Science, Technology, Engineering and Mathematics: Actions Needed to Better Assess the Federal Investment (Report No. GAO-18-290).

Hahn, H. & Lynn, B. (2017). An Innovative Program to Further the Careers of Women as Leaders in Engineering. *27th Annual INCOSE International Symposium*, Adelaide, Australia, July, 2017.

INCOSE. (n.d.). About Systems Engineering. Retrieved from www.incose.org/about-systems-engineering.

Inda, M., Rodriguez, C. & Pena, J. (2013). Gender differences in applying social cognitive career theory in engineering students. *Journal of Vocational Behavior*, 83, 346–355.

Islam, M., Kafle, S., Wong, A., Guillaume-Joseph, G. & Heart, D. (2017). Challenges and Opportunities for the New Generation of Systems Engineering Leaders. *27th Annual INCOSE International Symposium*, Adelaide, Australia, July, 2017.

Kasperovich, I. (2016). IEEE women in engineering (WIE). *IEEE Electromagnetic Compatibility Magazine*, 8(1), 40.

Kelley, M. & Bryan, K. (2018). Gendered perceptions of typical engineers across specialties for engineering majors. *Gender and Education*, 30(1), 22–44.

Kranov, A., DeBoer, J. & Abu-Lail, N. (2014). Factors affecting the educational and occupational trajectories of women in engineering in five comparative national settings. *International conference on interactive collaborative learning*, Dubai, UAE, December, 2014.

Lent, R., Brown, S. & Hackett, G. (1994). Towards a unifying social cognitive theory of career and academic interest, choice and performance. *Journal of Vocational Behavior*, 45, 79–122.

Lent R., Brown, S. & Hackett, G. (2000). Contextual supports and barriers to career choice: A social cognitive analysis. *Journal of Counseling Psychology*, 47(1), 36–49.

Leslie, L., McClure, G. & Oaxaca, R. (1998). Women and minorities in science and engineering. *Journal of Higher Education*, 69(3), 239–276.

National Academies Press (2006). To Recruit and Advance Women Students and Faculty in Science and Engineering. Retrieved from http://nap.edu/11624.

National Academies Press (2014). Career choices of female engineers. *A summary of a workshop*. Retrieved from http://nap.edu/18810.

National Center for Education Statistics Digest. Tables 290.2009, 317.2010, 318.30.2011–2016.

National Science Foundation (1989). *The State of Academic Science and Engineering*. Washington, DC: NSF.

Nilsson, L. (2015, May). Women who choose engineering [Opinion Column]. *New York Times*, Retrieved from www.nytimes.com/2015/05/06/opinion/women-who-choose-engineering.html.

Pollack, E. (2016). *The Only Woman in the Room: Why Science Is Still a Boys' Club*. Boston, MA: Beacon Press.

Robinson-Bryant, F. & Norman, N. (2017). Identifying critical systems thinking skills and opportunities across ABET programs. *Proceedings of the IISE annual conference*, Pittsburgh, PA, 2017.

Saavedra, L., Araujo, A., Manuel de Oliveira, J. & Stephens, C. (2014). Looking through glass walls: Women engineers in Portugal. *Women's Studies International Forum*, 45, 27–33.

Schwarzer, R. & Jerusalem, M. (1995). Generalized self-efficacy scale. In J. Weinman, S. Wright, M. Johnston (Eds.), *Measures in Health Psychology: A User's Portfolio. Causal and Control Beliefs* (pp. 35–37). Windsor, UK: NFER-NELSON.

Servon, L. & Visser, M. (2011). Progress hindered: the retention and advancement of women in science, engineering and technology careers. *Human Resource Management Journal*, 21(3), 272–284.

Singh, R., Fouad, N., Fitzpatrick, M., Liu, J., Cappeaert, K. & Figuereido, C. (2013). Stemming the tide: Predicting women engineers' intentions to leave. *Journal of Vocational Behavior*, 83, 281–294.

Systems Engineering Research Center- SERC (2018). Atlas 1.1: An Update to the Theory of Effective Systems Engineers (Report No. SERC-2018-TR-101-A). Stevens Institute of Technology. Retrieved from www.dtic.mil/dtic/tr/fulltext/u2/1046509.pdf.

Teyssier, J., Saenko, S., van der Marel, D. & Milinkovitch, M. (10 Mar 2015). Photonic crystals cause active colour change in chameleons. *Nature Communications*, 6, 6368.

Tharp, A. M. (2002). Career Development of Women Engineers: The Role of Self-Efficacy and Support Barriers (Master's Thesis). University of Minnesota.

Tietjen, J. (2017). *Engineering Women: Re-visioning Women's Scientific Achievements and Impacts*. Springer International Publishing.

United Nations Educational, Scientific and Cultural Organization-UNESCO (2017). Cracking the Code: Girls' and Women's Education in Science, Technology, Engineering and Mathematics (STEM). Retrieved from http://unesdoc.unesco.org/images/0025/002534/253479E.pdf.

U.S. Census (1890). Retrieved from www2.census.gov/prod2/decennial/documents/1890a_v1p2.zip.

U.S. Census (1990). . Retrieved from www2.census.gov/library/publications/decennial/1990/cp-s-1-1.pdf.

U.S. Census (2000). . Retrieved fromwww.census.gov/prod/cen2000/phc-1-1-pt1.pdf.

U.S. Census (2010). . Retrieved from https://factfinder.census.gov/faces/nav/jsf/pages/searchresults.xhtml?refresh=t.

Webster (2018). "Engineering" and "Industrial Engineering". In *Merriam-Webster's dictionary* (Online). Springfield, MA: Merriam-Webster.

6

Designing, Developing, and Deploying Integrated Lean Six Sigma Certification Programs in Support of Operational Excellence Initiatives

Jack Feng
Commercial Vehicle Group, Inc.

Scott Sink
The Ohio State University

Walt Garvin
Jabil Inc.

CONTENTS

6.1 Introduction

We combine our experiences to share how you can better prepare undergraduate (UG) industrial and systems engineers (ISEs) to be the major players in Operational Excellence Programs. We review how best-in-class Integrated Lean Six Sigma (ILSS) certification programs are designed and executed in industry and then map this to examples of how a few ISE departments are doing this in academia.

The editors for this book project state, "This book project arose out of a recognized need to explore the infrastructure, methods, and models for continuing and expanding the space for industrial and systems engineers (ISEs) with respect to academia and practice integration." We believe that this chapter fits perfectly. It is generally recognized that ISE missed a window of opportunity when Lean and Six Sigma were maturing and coming together. Recently, we have seen ISE begin to earn more recognition for its potential to contribute and even lead or drive these initiatives and corporate wide programs.

Interest in, and demand for, certifications that augment a BS degree in ISE (BSISE) or other degrees is growing and we believe has proven to add value to resumes and to enhance students' chances for finding great jobs in a shorter period of time. The value of a professional—ISE degree or other—that has been properly trained to effectively and efficiently tackle important process improvement projects is significant. Industry standards for a full-time Black Belt (BB) productivity (business impact), for example, would be 12–16 times their annual salary in benefits as an average benchmark. So, a certified process improvement specialist, earning $100,000 would be generating direct, indirect, and other benefits in excess of $1,000,000 annually. BBs are expected to manage portfolios of improvement projects and coach/lead Green Belt (GB) thus further expanding their impact to the value of the enterprise. Proper training and certification enhances the acceleration of the realization of these benefits to organizations.

Well-prepared, properly trained process improvement specialists add value easily and quickly. Poorly prepared specialists struggle, create negative experiences in the workforce, and cause yield loss in OpEx programs that can become fatal if not resolved.

Unfortunately, as this field has evolved over the past 20 years or so, gaining certification has gotten watered down, lost its way, succumbed to commercialization, and lessening of standards. The variation in options for getting certified today is wide and terms or operational definitions are confused, often purposefully. Certificate versus certification is an example of confusion and "watering down" of the process. The industry of certification has no governing body. American Society for Quality (ASQ) perhaps defines it for most but even their approach to certification is "light" in our view as evidenced by programs presented in this section. The real gold standards for certifications are held and upheld by a few high integrity consulting and training firms and curriculum providers and also largely by a growing set of companies that have developed solid, rigorous programs for certification to support outstanding Operational Excellence programs such as those shared in this section.

Our chapter focuses on defining and describing how these rigorous, high-quality certification programs are designed and are developed and deployed. We'll cover industry examples and one academic example. Our intent is to make the gold standard for ILSS certification a bit more explicit and to draw a line or Spec Limit on what rigor means relative to certification.

6.2 Example ILSS Deployment Programs

6.2.1 Commercial Vehicle Group's Lean 6-Sigma Programs

Incorporated in 2000, and Initial Public Offering (IPO) listed at the Nasdaq in 2004, Commercial Vehicle Group, Inc. (CVG) was formed through the acquisition of a number of companies around the vision of becoming a prominent supplier to integrated cab components and systems for commercial trucks, buses, construction, mining, and military vehicles, along with agricultural machines. The multiple brands CVG owns now date back to the 1930s and 1940s. With about 9,000 employees, CVG now has facilities in the United States, Mexico, the United Kingdom, China, Czech Republic, Ukraine, India, Australia, Thailand, and Belgium. In the past decade, CVG's sales ranged between $450M and $900M. CVG's key customers include Volvo, Daimler, PACCAR, Caterpillar, Deere, Navistar, VW, Tesla, Cummins, Oshkosh, and Yamaha. Overall, CVG's Operational Excellence (OpEx) Program has experienced three generations. The first generation was named Total Quality Production System or TQPS prior to 2013, the second generation was named CVG Operating System or CVGOS between 2013 and 2014, while the current third generation is named Operational Excellence/Lean 6-Sigma since the end of 2014 when Jack joined CVG. In this subsection, the term "Lean 6-Sigma" is CVG's program brand while you will see different expressions such as Lean Six Sigma and LeanSigma in this chapter.

6.2.1.1 CVG's Lean 6-Sigma Training Programs

Based on the same concept of developing our internal "Navy Seals" as outlined by Scott in section 6.2.2, CVG deems developing and growing our own Lean 6-Sigma belts as the first and most important step in our accelerated move toward Operational Excellence. Therefore, we have introduced five different Lean 6-Sigma belting programs (as shown in Table 6.1), and they are offered in English, Spanish, Chinese, Czech and Ukrainian languages. It is largely based on Caterpillar's former 4 weeks of 6-Sigma Define–Measure–Analyze–Improve–Control (DMAIC) BB program and 2 weeks of Lean BB program. These Caterpillar programs have been recognized as one of the top industry programs worldwide in different books, articles, and trade magazines such as Montgomery (2013, pp. 30–31) and George (2002, p. 227).

Jack was involved at other companies, at the corporate level, Asia Pacific regional level, as well as the divisional level in designing, updating, translating, or personally deploying these programs before he joined CVG. As a result of this and Jack's lifelong achievements both in industry and academia, he was elected Fellow of the Institute of Industrial and Systems Engineers (IISE) in 2018 and as one of five Lean Leaders of the Year in 2011. The latter honor is from the annual recognition by the China International Manufacturing Management Forum, China's annual Lean Forum. This recognition in 2011 was during Jack's international assignment leading the deployment of these programs to Caterpillar employees, suppliers, and dealers in Caterpillar's Asia Pacific Region.

Jack also developed a 40-h Lean Master Black Belt (MBB) training curriculum and taught the first four waves of classes for Caterpillar. In addition, Jack led the development and deployment of the Caterpillar Production System (CPS) Master Learning Center and the Train the Trainer Program.

TABLE 6.1

Current CVG Lean 6-Sigma Belting Programs

Name	BB	GB II	GB I	Yellow Belt II	Yellow Belt I
Purpose	All CI managers in large facilities must be full time and BB trained	Every small facility should have at least one GB II trained	Offered in Spanish to Mexican employees and suppliers	Support value stream level lean transformation or awareness	Support an area or station level kaizen event or awareness
Number of hours	160h in 4 weeks within 4 months	80h in 2 weeks of within 2 months	40h in 1 week	20h in multiple modular sessions	8–10h in multiple modular sessions
Trainers	Master BBs only	Master BBs + BBs	Master BBs + BBs	Facility BBs/GB s	Facility CI managers or BBs/GBs
Registration	Nomination by leaders	Nomination by leaders	Nomination by leaders	By registration, open to all CVG employees	By registration, open to all CVG employees
Language	English only	English, Chinese, and Spanish	English, Chinese, Spanish	English, Chinese, Spanish	English, Chinese, Spanish Czech Ukrainian
Supplier development		Open to suppliers by invitation			Open to executives of suppliers by invitation
Curriculum Owner	Global OpEx Team with expertise in English, Chinese, and Spanish				

This includes the 1-week CPS Assembly Learning, the 6-week CPS Machining Learning, and 7-week CPS Welding Learning programs. These programs were developed based on the Training Within Industry model he recommended to Caterpillar as a lean consultant prior to becoming a full-time Caterpillar employee.

The format of CVG's program is built upon the best practices in the industry. For example, a BB class is offered over 4 weeks within 4 consecutive months at different facility locations. This model is echoed in the Jabil model in the next subsection. Jack and his team found that offering the class at different facilities with a classic waste walk or Gemma walk tour to these different operations would give the students an enterprise view; otherwise, their view of CVG would be limited to their respective location and those people around them. At the same time, this gives these future leaders exposure to different product lines, different operations, and local culture and foods. Due to the size of our company, CVG offers the last 2 weeks of our BB class in North America only, while the GB classes are also offered in China, Mexico and the United Kingdom in Chinese, English, and Spanish.

Each BB or GB student must bring to class a lean kaizen or 6-Sigma project approved by their local leader. This follows the best practice outlined by Harry and Schroeder (2000) who pioneered the Motorola, Honeywell, and GE's Six Sigma Programs. Before the nomination of any classes, redundant training sessions are offered to make sure the facility manager, business manager, or equivalent leader understands how to nominate candidates, how to select the best project for the candidate, and how to become a good project sponsor. A sample list of BB nomination criteria is shown in Table 6.2. Figure 6.1 shows the process of our project selection based on pioneers from Dow Chemical and GE (Snee and Hoerl 2003), while Table 6.3 is a sample list of tactical project selection criteria.

CVG's BB graduation week is always offered at its New Albany, Ohio headquarters location so that the corporate senior executives, from the CEO to product line or business line heads, can attend the graduation and the signature Gallery Walk. The Gallery Walk offers the unique features. One, every BB student posts his/her one page story board on 36"×48" paper around the classroom and the adjacent lounge room. Two, after a brief introduction of all BB students, MBBs, and senior leaders, during the Gallery Walk event, the senior leaders are invited to randomly walk through the story board. The Project Leader or BB stands in front of his or her "stall" ready to tell his/her story to the senior executives. This event not only gives individual recognition of these future leaders but also provides an opportunity for our executives to learn what projects are being worked on or what problems are being addressed in each facility or business unit. Furthermore, this helps broaden the skill set of these future leaders by speaking in front of an audience of senior level executives.

TABLE 6.2

A Sample List of BB Candidate Nomination Criteria

Lean 6-Sigma GB/BB Nomination Criteria	Reason or Rationale
1. The candidate should be full-time CVG employee by the starting date of class	To ensure sufficient time to identify a project and complete project charter creation and approval
2. Each facility has minimum one GB; each facility with $20M + annual revenue has at least one BB; every $20M annual revenue has one BB, or target four BBs for a facility with $80M revenue	To ensure the appropriate density and people capability in delivering cost out target for the facility or Business Unit
3. The candidate has played prior leadership role(s)	Precondition to being a good project leader and for Criterion 5
4. The candidate has a 4-year college degree or equivalent post high school education	To ensure the candidate will not fail the class training with about one-third of contents on statistics and statistical methods
5. The candidate has a great potential to become a leader within a facility or supporting functional area within 1–3 years of graduation	To ensure we are investing into the right talent given the investment cost and limited resources to provide the training
6. The candidate has at least 3 years of combined full-time work experience	To ensure good motivation, successful project execution, and serve as a precondition to criteria 3 and 5
7. Upon completion of training, the GB candidate is desired to stay full-time employment within CVG for at least 1 year; the BB candidate needs to serve CVG for 2 years after graduation; otherwise, the employee needs to pay back the CVG investment on the pro-rate basis.	Reasonable payback of our return on investment. Estimated weekly based on market value or opportunity cost with comparable programs. Consistent with corporate tuition and relocation reimbursement policies.

While Table 6.4 summarizes major contents of CVG's BB/GB programs, we have additional features as outlined next. (i) We have the **Train the Trainer** module in each week. Under this module, all students are required to present their project charter in the first week or project progress in the subsequent weeks in 4 min leaving 1 min for coaching and Q and A. In each week, all students are also required to teach back a module in 4 min leaving 1 min for coaching and Q and A. In week 4, we also offer the Train the Trainer module to train and coach the BBs how to facilitate the **Race Car Lean Assembly Game** so when they go back to their facility or business unit, they are capable of facilitating the game in support of each of the local Yellow Belt (YB) I or YB II classes. Figure 6.2 demonstrates a shift of assemblers engaged in the Race Car Lean Assembly Simulation Game. (ii) We use the

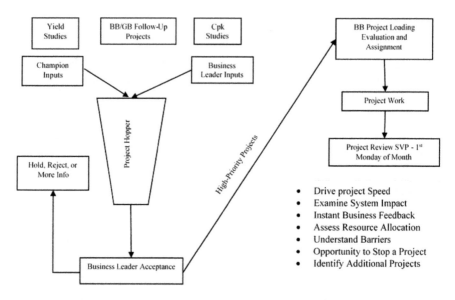

FIGURE 6.1
BB/GB project selection process (Snee and Hoerl 2003, p. 123, with permission from Pearson).

TABLE 6.3

A Sample List of Detailed Lean or 6-Sigma Project Selection Criteria

6-Sigma Project	Cross-functional teams, involving suppliers, or customers, or other sister facilities; true root causes and solutions unknown; top priority in support of corporate short- or long-term strategy and cost out; normally takes 2–6 months.
Lean or Kaizen Project	Within one facility or one functional team; root causes and solutions less unknown; normally takes 1–14 weeks.
Quick Win or Just Do It Project	Root causes and solutions known; just need to assign resources to get it done. A part of daily job. Does not need a GB or BB. No innovation but implementation tedious; time duration linearly

Catapult Game as a hands-on tool to teach students a range of skills such as charter creation, analysis of data, design of experiments, and the identification and minimization of variations. Figure 6.3 illustrates a team engaged in the catapult competition—within limited space in this case. (iii) From week 2 on, we use **Minitab** as the tool to teach and practice various 6-Sigma tools. (iv) We have **group exercises,** and group report backs are included every day to foster team building, group dynamics, and the practice of various Lean and 6-Sigma processes and tools. Teamwork and group dynamics are gaps of education programs in some countries where CVG operates.

TABLE 6.4

Time and Contents Allocation of CVG's BB and GB Programs

Topics	% of Time Spent
Eight wastes, 20 lean guiding principles plus other lean tools: VSM and VST, RIW (Rapid Improvement Workshop), BIQ, TPM, OEE, standard work, 5S, QCO, visual factory, flow, pull, level, metrics cascade, etc.	40
Six Sigma statistical tools/methods: DMAIC, 8D report, SIPOC, MGPP, 5 whys, fishbone diagram, FMEA, 80/20 rules, control charts, ANOVA, DOE, hypothesis tests, Pugh Matrix, generating random ideas, etc.	40
Leadership development: Facilitation skills, change management, project planning and management, building an effective team, financial benefits estimate, risk assessment and mitigation, presentation skills, teaching skills, and practices	20

FIGURE 6.2
Race car lean assembly simulation game (Photographer: Kerry Howery, CVG MBB).

6.2.1.2 CVG's Lean 6-Sigma Deployment Model

In its current deployment model designed by Jack and approved by the President/CEO, Jack, as the Vice President of Operational Excellence and CVG Digital, reports to the CEO directly. Three Lean 6-Sigma Deployment Managers (DMs)/MBBs report directly to Jack with each in charge of a product line and responsible for one senior operations executive. A fourth Deployment Manager serves the various corporate functions. Among the 20+ manufacturing facilities across different parts of the world, each facility with sales and revenue no less than $20M is required to have a full-time Continuous Improvement (CI) Manager reporting to the respective facility manager but

FIGURE 6.3
Hands-on catapult game to minimize variations (Photographer: Kerry Howery, CVG MBB).

with a dotted line reporting to the respective product line's DM/MBB. CVG is a relatively small organization as opposed to the majority of our customers. It is difficult for us to justify a separate DM role and MBB role, so the leaders we recruited to fill these roles previously held various functional leadership roles and had at least 15+ years of Lean BB, 6-Sigma BB, and CI working experience, and some of them were MBBs well before joining CVG.

Similarly, due to the relatively small size of the various CVG facilities, none of our BBs or GBs are full time, as opposed to our large customers such as Caterpillar, PACCAR, Deere, and Cummins. The only full-time BBs or GBs are those in the facility CI Manager roles. The CI Manager in a large facility must be BB trained, and those in relatively small facilities must be GB trained. Tables 6.5 and 6.6 summarize the additional commonalities and differences, respectively, in deploying OpEx/Lean 6-Sigma to small or large organizations (Feng 2016).

As detailed in Scott's section 6.2.2, a rolling 5-year plan is created to highlight the MBB/BB/GB density, bottom-line savings, focus, and owners based on sales projection and some turnover assumptions.

To help cascade the corporate Hoshin Plan to the smallest possible manufacturing value stream level, we have standardized the **OpEx Metrics Board** at the facility/business unit level and the value stream levels. Figure 6.4 illustrates the facility level OpEx Metrics Board with the categories of People, Quality, Velocity, and Cost (PQVC). From left to right, it also follows the PDCA (Plan–Do–Check–Act) cycle. Some metrics are common among facilities such as safety and CI suggestions per employee under People, internal first-pass yield and customer defective Part Per

TABLE 6.5

Major Differences in Deploying OpEx/Lean 6-Sigma between Small and Large Organizations

Key Areas	Small	Large MNCs
Organization: DM and MBB roles	Same person	Different people
Organization: Key process owners	Part time or not available	Full time
Resources: Facility CI manager role	Only major facilities could afford an FT role	Most facilities have an FT role
Resources: BB role	Part-time except some FT CI manager roles	Active BBs are all FT
Curriculum	Purchase/could not afford the huge spend and time	Afford big spend and time due to critical mass
Infrastructure	No FT staff to create and support	FT staff to create and support
Coaching: MBB to BB + GB ratio	1:~30+ ratio common	About 1:15 ratio
Suppliers deployment	No dedicated resources	Dedicated resources
Training/certification logistics support	No FT staff	FT staff
Velocity of deployment	Flexible and usually faster	More disciplined and usually slower

TABLE 6.6

Commonalties in Deploying OpEx/Lean 6-Sigma between Small and Large Organizations

Key Areas
Top leadership team commitment: Must
Board of directors commitment: Must
Passionate, capable, and energetic champion(s): Default
Top down and bottom up engagement: Must
Always emphasize "Control" or sustain in DMAIC
Employees engagement, not just employees involvement, is key to sustain or control
Step by step with multi-generation plan
Drive for and document internal financial results
Focus on customer quality and delivery
Always apply the 80/20 rules in prioritizing projects and allocating resources
Always apply the PDCA cycle in project/process management
Start internal then expand to suppliers base
Focus on manufacturing operations first then support or transactional operations (for manufacturing organization)

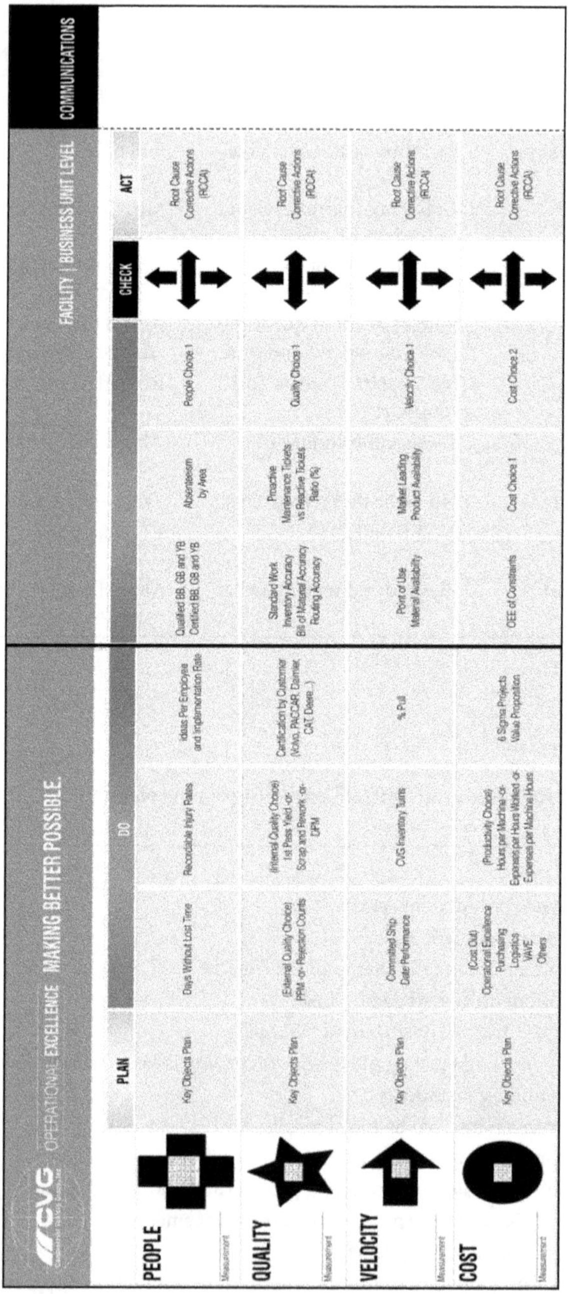

FIGURE 6.4
Standard facility/factory-level OpEx metrics board.

Million (PPM) under Quality, on-time-shipping performance and inventory turns under Velocity, and cost out and OEE at constraints under Cost. A number of facilities with a full-time facility CI Manager available have been using this as one of the regular process improvement dialog tools, along with the use of our standard **CI Board** as shown in Figure 6.5. All the CI suggestions are organized into the PQVC OpEx Metrics categories by the local CI Manager.

The standard CI Board is designed based on multiple visual management concepts such as the color of the two types of T-card: the dark gray color card is for safety-related CI suggestions only, while the light gray color card is for all other CI suggestions. The small dough attached to each CI card with different colors is used to represent different departments in charge of closing or implementing the CI suggestions. Major columns from left to right indicate different stages of the process of the card: New, Assigned, In Process, Past Due, or Closed.

Tracking and reporting financial benefits is a common critical success factor for any OpEx Program. We have developed a web portal to track and report the above benefits into three categories by MBB/BB/GB and by wave of classes: Level 1 savings—bottom-line or hard savings; Level 2 savings—OPACC savings (Operating Profit After Capital Charge following the 1929 DuPont model), and Level 3 saving—soft savings. Within CVG, the Category 1 or hard savings must be a dollar-to-dollar match to the savings reported monthly in our profit and loss statement, while the OPACC saving from either of the following three operating working capital categories will give

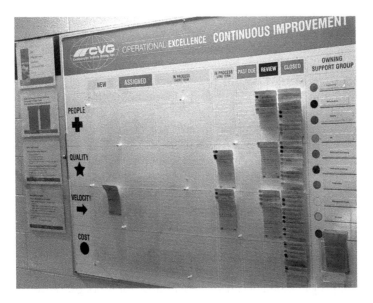

FIGURE 6.5
Standard CI board for visual management.

the BB or GB credit toward certification but not reported as hard savings: (i) inventory reduction, (ii) accounts receivable increase, and (iii) accounts payable decrease. Every month, the first two categories of benefits by MBB or Coach are part of the monthly Hoshin review with the CEO and his leadership team.

In summary, over the past 3 years of CVG's journey in deploying the new OpEx/Lean 6-Sigma model, we have improved the entire corporation's quality and inventory turns to a statistically different level. The cost savings as percentage of sales from Lean 6-Sigma projects, led by our MBBs, BBs, and GBs, have become increasingly larger in the corporation's total cost-out tracker, and they have become increasingly more predictable and reliable as well. By the end of 2018 about 23% of all CVG employees have been BB, GB, or YB trained.

Furthermore, among our worldwide suppliers base, CVG has trained about 40 supplier BBs or GBs, in addition to about 100 YB trained supplier executives such as Chairman, President, VP for Operations, and Director of Quality. Several suppliers have started to deploy our OpEx model to their operations in addition to pockets of excellence in working on Lean kaizen or 6-Sigma projects. As a consequence, the OpEx program and its contribution to our margin enhancement have been recognized in CVG's quarterly and annual reports, and in various investor relation road shows by the company's CEO and CFO.

For a more detailed historical account of CVG's OpEx Deployment process, refer to a phone interview published in February with the Executive Platforms (Feng 2017b). Recently, Jack has completed several speaking engagements to various industry groups. The content included success stories and case studies (Feng 2016, 2017a). One of the success stories describes Overall Equipment Effectiveness **(OEE) improvement** resulting in avoiding millions of dollars in capital investment. Another focus is the deployment of the **Manufacturing Plan for Every Part (M.PFEP)** template and process in support of an improved Sales and Operations Planning (S&OP) process. This change has dramatically improved inventory turns in a few large facilities while maintaining or improving on-time-delivery performance and premium freight costs. A third discussion centered on the implementation of the **Built-in-Quality** (BIQ) recipe leading to significant reduction of customer PPM, increase of first-pass yield, and reduction in the end of line rework resulting in improved customer recognition.

How **change and risk management** are handled is one generally recognized requirement for sustained success in any OpEx programs at both the enterprise macro-level and the project microlevel. That is why our BB/GB Programs dedicate 20% of our time and content on leadership development. One commonly cited survey (Prosci 2000) indicates over 50% of the time, a program or project failed because the people factor is not properly addressed: lack of upper management commitment or lack of engaging frontline associates. Therefore, we teach and preach the following recipe for success.

It means people engagement is at least as important as the technical solution of a project or a program.

$$\text{Success} = \text{Quality of Solution}\,(\text{Qs}) \times \text{Acceptance of the Solution}\,(\text{A})$$

Jack believes an important factor in CVG's journey toward Operational Excellence to be the firm commitment of CVG's two successive CEOs and its CFO. The prior CEO Mr. Rich Lavin retired from Caterpillar as one of its five Group Presidents. As an officer, Mr. Lavin was part of Caterpillar's journey in starting the Six Sigma deployment in 2001 and the Six Sigma-based CPS Deployment in 2005. The Caterpillar Japanese operations were under Mr. Lavin's Profit & Loss portfolio as well. His faith in the proven benefits from Caterpillar's experiences and success stories led him to give Jack the freedom to jump start CVG's third generation OpEx Program.

As an Industrial Engineering (IE) graduate from Purdue with an MBA from Harvard, CVG's current CEO, Mr. Patrick Miller, has a strong process background rooted in lean principals developed while working at ArvinMeritor and other tier one automotive suppliers. These experiences allowed close partnerships with all of the top automotive Original Equipment Manufacturers (OEMs) including Toyota and Honda generating successful deployment of many of the tools encompassed in the CVG program. As a result, Jack and his team benefit from the unwavering commitment by CVG's top executive which aids in accelerating the momentum of CVG's OpEx journey. CVG's CFO, Mr. Tim Trenary, has also been committed and supportive of our OpEx journey after witnessing CVG's first year success and financial bottom-line improvement in 2015 and then in the successive years. Consequently, the top leadership team's commitment which is required to make any OpEx Program successful and sustainable as outlined in Table 6.6 has not been an obstacle for Jack and his team at CVG.

6.2.1.3 CVG's Lean 6-Sigma Certification Process/Programs

It is difficult to make any changes in people's behavior. It is even more difficult to sustain these changes. In order to help achieve this, we have introduced several structural change management processes. One of them is the **BB/GB Certification Process**. Table 6.7 is a sample list of our BB and GB Certification Criteria. We consider this list of criteria to be more stringent than that from the Lean or Six Sigma BB or GB certification programs offered by ASQ and IISE. Our process requires training or facilitation hours, financial savings, and a completion of minimum of two to three projects to become certified.

At the end of 2015, we introduced the **CVG President's (OpEx) Award** at the Value Stream Level. Six criteria in the PQVC categories are used to assess a value stream based on five levels as shown in Table 6.8. In March 2017, CVG presented the two inaugural President's Awards including the Best Lean Value Stream, which was awarded to a U.S. facility, and the Most Improved

TABLE 6.7

A Sample List of BB/GB Certification Criteria

		BB	GB
Criteria	1	Successful completion of the 4 weeks of training	Successful completion of the 2 weeks of training
	2	Realized $150K+ green dollar saving	Realized $80K+ green dollar saving
	3	Successfully led and completion of minimum four 6-Sigma projects/RIWs	Successfully led and completion of minimum two 6-Sigma projects/RIWs. If the GB participated in two BB projects, it will be counted as leading one
	4	Provided 40+ hours of Lean 6-Sigma training	Provided 20+ hours of Lean 6-Sigma training
	5	For BBs in operations, the 40 h of training must include one facilitation of Race Car Simulation and one facilitation of an RIW	For GBs in operations, the 20 h of training must include one assistant facilitation of Race Car Simulation AND one facilitation of an RIW
	6	Constant performance rating PR3 or above	Constant performance rating PR3 or above
Reward	1	Promotion to BB 2 and eligible for promotion to a major department head position within a facility or equivalent in a nonmanufacturing organization	Credit counts toward future BB certification
	2	A one-time bonus 1% of value prop in green dollar saving and OPACC contribution	A one-time bonus 1% of value prop in green dollar saving and OPACC contribution
	3	CVG certificate	CVG certificate
	4	Candidate for future master BB	Nomination to the next BB class

TABLE 6.8

The CVG President's Award Assessment Criteria

Criterion Category and Weight (%)	Criterion Name	Criterion Weight (%)
People (30)	OSHA recordable injuries	30
Quality (30)	First pass yield %	15
	External PPM	15
Velocity (20)	Schedule attainment %	20
Cost (20)	Efficiency % or labor hours/unit	10
	Downtime %	10

Value Stream, which was awarded to an international facility. Since then, six value streams have win either of the two awards.

While the President's Award served well as a **voluntary participation** program to drive grass root-level cultural and behavior changes, we subsequently rolled out the **mandatory Lean Value Stream Certification Program** at the end of 2017 in order for us to maintain the momentum and for our OpEx Program to "Cross the Chasm" based on the concept presented in Moore (2001). Each facility must certify at least one value stream as its Model Value Stream following the Toyota Production System model deployed to suppliers. Given the density of our BBs, GBs, and YBs, the minimal level of CI Managers available, and the increasing desire of moving toward Operational Excellence in every job we do, CVG believes our conditions are mature enough to roll out this mandatory program at this stage of our OpEx journey. We have identified seven metrics among 16 total metrics as mandatory. The remaining nine metrics are optional but a facility must choose at least three of them to certify a Manufacturing Value Stream.

Following the U.S. model for the Malcolm Baldrige National Quality Award, we also assess six key lean support or enabling processes in addition to the above ten metrics focusing on improvement results only. This list of six Lean support processes includes Safety plus 5S (6S), BIQ, OEE, Pull, Level Loading (microlevel S&OP), and Supplier Development & Collaboration. Following the above CVG President's Award scoring, each metric or support process is certified into five levels as outlined in Table 6.9. An average of 4 or a total score of 52 is required to certify a value stream as a Lean Value Stream: 40 points from the ten metrics and 12 points from three of the six support processes—Safety plus 5S is mandatory. The highest scored value stream-certified facility and most improved value stream-certified facility will be awarded the 2018 President's Award for Lean Value Stream and the Most Improved Value Stream Award, respectively.

TABLE 6.9

Operational Definition of Five-Level Lean Assessment

Level	Delineating Differences in Operational Definition
1	No proactive work has started in the metric
2	*Policy has been established on paper*, but no implementation has started
3	Implementation has started for less than 3 months, *but target has not been attained*
4	Structural lean and 6-Sigma tools were applied to improve the process, product or service, and target has achieved and maintained for *at least 3 months* in the fiscal year
5	Structural lean and 6-Sigma tools were applied to improve the process, product or service, and target has achieved and maintained for *at least 6 months* in the fiscal year

To encourage the same behavior changes in administrative or transactional areas or functional teams, we also introduced the President's Award for non-Manufacturing Value Streams in 2018. The 2018 President's Award was given to the cross-functional teams that have developed and validated our accounts receivable process piloted at one of our Mexican facilities.

CVG rolled out a comprehensive CVG Digital Strategy in the beginning of 2017. CVG Digital is a required part of the Lean Value Stream Certification process. For example, each value stream to be certified needs to install the corporate recommended standard real time bingo board or production dashboard based on bar code scanning plus the standard Andon stack lights in support of real-time data collection and display to ensure faster root cause analysis and corrective action planning. For example, our standard bingo board collects and displays real-time data about OEE, first-pass yield, hourly actual versus hourly target, accumulative actual versus accumulative target, and cycle time versus time remaining within cycle. Our four-stack Andon lights and the keypad system records real-time downtime data based on the nine OEE loses and automatically sends text and e-mail alerts to the responsible party or parties for fixing the unpredicted production problems such as machine breakdown, waiting for material, waiting for associate, and stopping to fix quality issues.

6.2.2 Global Life Sciences Deployment of ILSS and Operational Excellence

GLS, Inc. (disguised) was an organization comprised of four Revenue-Generating Business Units and one Enterprise Shared Services Organization. The organization provided Medical Isotopes, Analytic Technologies, Pharma Services (Early- and Late-Stage Drug Testing), and Laboratory Services/Testing. Revenues of approximately $1.2B and 5,500 employees in 2004 time frame.

The Board of Directors had approved a major Transformation Initiative that involved formalizing an Enterprise Shared Services Business Unit as well as major organic and inorganic growth strategies. A new CEO was brought in to lead the transformation. A President for Enterprise Shared Services was hired, and that person recruited Scott Sink for the role of VP Business Process Improvement.

At one of the new CEO's initial top 100 meetings, he unveiled his "Platform" (strategy) and "Planks" in his strategy, and you can see in Figure 6.6 that Operational Excellence, ILSS, and even ISE were well positioned in the larger transformation. With that as the launch point, the author of this section joined GLS, Inc. in the VP of Business Process Improvement Role and was charged with Enterprise-Wide Deployment of Business Process Improvement (BPI) (to include reengineering) and reported to the President of Enterprise Services with dotted line to the CEO.

	Process	Outcomes
Business Performance Reviews	• Weekly EMT teleconferences • Monthly business reviews • Disciplined annual plans	• Action oriented decision making • Tighter accountability • Customer responsive
Talent Management	• Biannual talent reviews • New executive compensation plan	• Better understanding of "A" performers; enriched career path • Expansion of variable compensation opportunity • Alignment of shareholder and management incentives
Customer/ Competition/ Capital	• Business unit/Corporate strategy • Detailed industry analysis • Customer value led process	• Longer range growth agenda • Focused R&D investments • Capital matched to growth
Operational Excellence	**Lean Sigma Roadmaps and Toolkit** **Compliance Programs (EHS, Quality, etc.)** **LeanSigma Practitioner Development** **Balanced Improvement Portfolios**	**Standard approach across the Enterprise** **Building global quality competitiveness, productivity improvement, process and cost efficiency, compliance and assurance** **Simplify processes** **Customer responsive**

FIGURE 6.6
GLS transformation strategy.

A team comprised of seasoned BBs and a few very seasoned ISEs was assembled rapidly. Over the span of a year, MBBs, one from the Pharmaceutical Industry and one ISE with much industry experience, were added to take on Business Unit Deployment Leader roles in two of our Revenue-Generating Business Units. The term "we" going forward will refer to the BPI VP and the BPI Design and Development Team (DDT) which included Deployment Leaders in each of the business units and deployed belts that the program either hired or trained and certified.

The goal was to create a world-class, best-in-class Operational Excellence Program that would drive direct, indirect, and other benefits in excess of 2% annually, a sustainable Return on Investment (ROI) of 6:1. The end game as the Executive Vice President of Human Resources called it was to create a culture and process/system that would drive improvements in what we did and how we did it, across the entire business. We had specific Enterprise Value targets for each business unit and shared service support units that were linked to the process improvement project portfolio each year.

6.2.2.1 End Game

At one of the early strategy development sessions we held with Senior Leadership, the EVP of HR asked us what the end game was to all

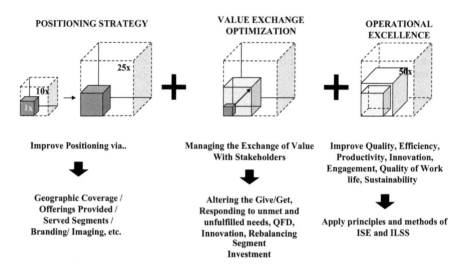

FIGURE 6.7
Three fundamental ways to grow enterprise franchise and shareholder value.

of this. He was a retired Navy Pilot, disciplined, pragmatic, focused, and intentional. Our response was to grow Enterprise Value for our stakeholders (shareholders, employees, customers, suppliers).

Figure 6.7 depicts a conceptual model/framework for understanding Enterprise Value. We defined Enterprise Value for the top team as a "cube," *x*-axis = the number of customer relationships, *y*-axis = the profitability of those relationships, and *z*-axis = the stickiness and durability of those relationships. The leaders in the room liked that model, it fit with the CEO's Platform and Planks, and the EVP of HR bought the answer.

Note that ISEs most typically "play" on the right side of this model, productivity, efficiency, quality, and the quality of work life. In our program at GLS, Intl., BPI (ISE) played in all three domains of improvement activity which gave our unit much more clout and kept us an integral part of the business strategy.

6.2.2.2 The Causality between an OpEx (ILSS) Program and the End Game

Figure 6.8 depicts the causal connections for ILSS programs and Enterprise Value. The Rath and Strong's Integrated LeanSigma Road Map is depicted in the bottom-left corner, that's DMAIC. Above that the "icon/image" reflects Continuous Improvement (2-s lean), kaizen events, DMAIC, Design for Integrated Six Sigma (DFILSS), and the various types of improvement initiatives/projects. The lower middle/right image is the Deloitte Touche Enterprise Value Map.

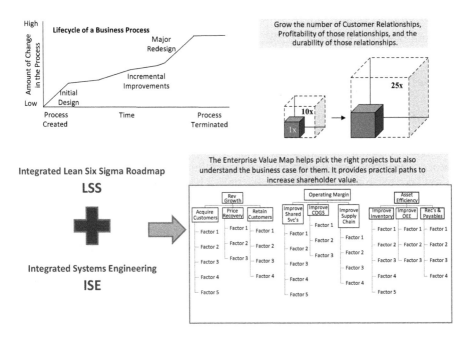

FIGURE 6.8
Causal connections from ILSS roadmaps, enterprise value map, and franchise value cube.

We utilized the Enterprise Value Map in our 2-day White Belt Training for the top 250 leaders in the organization. We found it very useful to help leadership think about selecting the right projects to drive business value. It was very useful in understanding causal connections between the ILSS projects and how value is created throughout the organization with the total portfolio of process improvement projects. We utilized this logic diagram as the guide for our project selection process. They experienced a 6-h "lean physical simulation (ProdSim-Sticklebrick)" (Prod Sim 2018) that allowed them to internalize the principles, methods, and potential for Lean and Six Sigma. We also did short Nominal Group Technique sessions to start to build the portfolio of improvement projects in each unit. Once the units were able to gain a measure of consensus of where the opportunities were, then they felt the need for the training and certification. Hence, we were able to create a pull for the certification program, which is exactly what we wanted.

We knew it was important to get alignment on the end game and the strategy, instrumentality of ILSS to the end game, and these rapidly deployed White Belt Sessions were invaluable in terms of socializing the need for and value for ILSS Deployment.

In all of our training (White Belt, GB, BB), we kept stressing the causal connection between the DMAIC methodology and how value is created for customers, employees, the business, and shareholders. Figure 6.13 reflects the constructs that we worked to get our team members to understand and internalize and ultimately drive.

6.2.2.3 Instrumentality of Training "Navy Seal" Quality Process Improvement Specialists

One of the key components of our Program at GLS, Intl. was our Process Improvement Specialist (Belts) training and development (and certification) program. Research and considerable experience has shown that driving process improvement with skilled specialists is significantly superior to allowing teams muddle through themselves even if those teams have had process improvement training. We did extensive benchmarking to organizations that were 5–7 years into their OpEx programs. We wanted to learn from people who were 5 years ahead of us not just from organizations such as GE and Motorola and Toyota who have been at this a very long time.

We became convinced that a formal ILSS training and development program was important to the success of our transformation strategy. We learned that skillful, disciplined ISEs and ILSS "belts" were able to drive improvement project success faster and better than teams with less skilled "facilitators." GLS, Intl. leadership supported the development of a team of "Navy Seal" quality process improvement specialists. We let the business units, such as Revenue-Generating as well as Shared Service Support units, IT, Finance, HR, Supply Chain Management (SCM), and Legal, "pull" for training. The business units development 3-year improvement plans that were a combination of challenge targets from the overall business as well as grass roots, voice of employee, and needs/initiatives. We applied many Hoshin Kanri-type methods as we developed our improvement plans and the improvement project portfolios.

The project portfolios at the BU level is what drove our training and development and "Belt" staffing requirements. We employed a combination of full-time staff in BPI along with outside hired BBs and MBBs to pump prime the launch. Then we systematically started to schedule and deliver GB and BB blended training and grew our team of process improvement specialists over the next 3 years.

6.2.2.4 Requirements for Design, Development, and Deployment of our ILSS Competency Development Program

Initially, we employed the services of BMGi, Breakthrough Management Group. Their team did a great job helping us formulate the initial strategy. They assisted us in developing our overall e-handbook, the Program Definition Document that was posted on the company intranet. This was developed by locking the top 20 level 2 leaders in the organization in a room and banging out the policies and procedures and requirements for the over-all BPI program.

We fleshed out things such as the following:

- Full-time versus part-time balance
- How teams and belts, project leaders would get incented and retained (how avoid investing in belts only to have them stolen away)
- Length of training, certification requirements, coaching, oversight, etc.
- Portfolio, project selection process
- Financial reporting, business case protocol, what we called Finance 601
- Infrastructure for the BPI program
  - We created a shadow infrastructure, Value Stream Owners were named, process owners, deployment leaders in every business unit, etc.
  - We mapped a matrix of who was accountable for running/administering the business (A-work) to what the accountabilities were for who would improve what we did and how we did it, build the business (B-work).

We followed the guidance of BMGi but also applied some Quality Function Deployment (QFD)-type discipline to our design and development process.

Given the short space allocated for this case study, the most important thing to share is the actual training design. We migrated from the traditional "death by Powerpoint" GB and BB training that is common in business and industry to a Blended Training Model. Our ongoing benchmarking activities led us to a company called Moresteam (Sink 2016). They primarily provide blended training program support, web-based training curriculum in support of ILSS certifications. We adopted their blended training model and found it to be a more effective and efficient way to deliver the training and develop the specialists and teams.

6.2.2.5 "Fronts" to Manage over Time

At the outset, we conceived of the BPI capability development for GLS, Intl. as a program, a 3- to 5-year migration strategy and plan. We utilized a "Grand Strategy Planning Model" (see Figure 6.9) and fleshed out all the various fronts we knew we had to manage. So, the Process Improvement Specialist Development initiative was developed and executed in the context of the much larger program. This was essential in our view to the success and sustainability of this work.

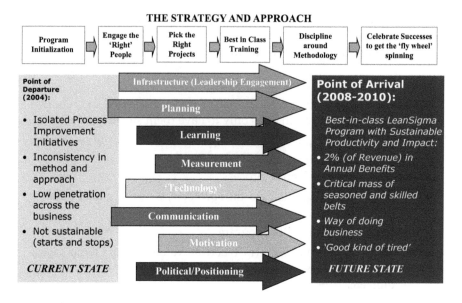

FIGURE 6.9
The grand strategy migration plan at GLS.

6.2.2.6 Design and Development Strategy and 3-Year Migration Summary

We created a "directional" 5-year plan for the program based on our learnings from the benchmarking we did. We integrated the 5-year Operations Excellence plan with the evolving enterprise business plans to ensure they were complimentary. As you can see, this included forecasts/estimates for benefits, investments, benefit-to-investment ratio targets, number of trained specialists by type, etc. (see Figure 6.10).

6.2.2.6.1 Close Out

This case represents a successful start-up, launch, and early maturation of an ILSS/OpEx Program. It was a systematic, well-engineered design and development project that was well integrated into the larger organizations strategies for growth. As you saw, the BPI program was well positioned in the business strategy. A lot of time and energy was invested in gaining alignment and buy-in from all the BU Leadership Teams. We completed 2-day White Belt training for over 40 teams, starting with the top teams in every business unit. That single intervention socialized ILSS and process improvement to well over 1,000 of the 5,500 employees across the entire enterprise.

At mid-2007 point, this program was well entrenched into the organization and was sustainable and a productive driver of improvement for each of the business units. One of the business units actually ended up getting sold off to generate capital for investment in more strategic areas of the enterprise.

	2005-06	2007	2008	2009	2010	2011	2012
Hard/Total Benefits	$6M/ $10M	$10M/ $16M	$20M/ $30M	$30M/ 45M			
Gross Revenue	$1B	$1B					
Benefit/ Revenue	.6%	1.6%	2%	2.2%			
Investment	~$3M	~$6M	~$7M	~$7M			
Benefit/ Investment	<1:1 (2:1)	2:1 (3:1)	3:1	6:1	6:1	6:1	6:1
Belts: M/SBB	3 (1:4 BB)	4 (1:8) BB	5 (1:7) BB	6 (1:10) BB	1:10	1:10	1:10
FT Belts	12 (.2%)	24 (.4%)	40 (1%)	60 (1%)	1%	1%	1%
PT Belts	51 (.9%)	130 (2.5%)	220 (4%)	300 (6%)	350 (7%)	500 (10%)	10%

FIGURE 6.10
The directional 5-year business case and model for OpEx deployment.

The CEO shared with me that the ILSS program in that Lab Services Business probably was a big factor in at least $300M of the end value gained from that sale. That in and of itself is clear evidence that we had, in fact, contributed to the "end game" the EVP of HR asked me about early in the program.

6.2.3 How to Develop Successful ILSS Programs in ISE Departments

6.2.3.1 Ohio State University's ISE ILSS Certification Program Launch and Evolution

There are many paths to the same end as they say. In general, ISE was late and ineffective at being a key player in the Total Quality Management (TQM) and ILSS "movement" that occurred from about 1990 and continues today. ISE departments create about 4,000 UG ISE graduates each year. The bulk of those BSISEs coming out will have little to no exposure to ILSS. However, in the past 10 years, ISE departments have adopted three approaches to addressing the ILSS/Business Process Improvement popularity and demand:

1. Ignore it, leave it to business and industry and other education and training "vendors" to handle.
2. "Buy" a solution—as example, take advantage of IISE's GB training offering and have them come in and do special training for students interested.

3. "Make" a solution, leverage a faculty member who has knowledge, skill, interest in this (often this would be the quality-focused faculty member) OR bring in a lecturer from industry who would do this part time perhaps.

We will speak to how a department might do the third approach.

6.2.3.2 Background

In June 2007, the Department Head of ISE at Ohio State University (OSU) and I met, and we discussed the possibility of adding an ILSS certification program to the ISE Curriculum at OSU. She shared insights she had gained from Senior Exit Interviews and from conversations she had been having with selected alums and employers. An overriding theme was that the students valued the course work but didn't see how things all fit together, and our Senior Design Capstone Course, 1 semester, wasn't effective at achieving that. She also felt that our UGs were under invested in, that there was a strong focus on the graduate students and research but that we needed to do something for our UGs.

I shared my experiences in industry with new ISE hires, what was working and what was lacking. We discussed knowledge and skill sets that we wanted Graduating Seniors to have; what we felt hiring employers wanted; and then compared those habits, traits, knowledge, and skill sets to what we felt the "reality" was, recognizing that there are "segments" of students, there is variation across students. This discussion and analysis created the "requirements" for the program we designed.

To make this section of the chapter efficient (three pages or less), I'll cut to the quick and say that basically we tailored the program from Global Life Sciences Intl to fit into the Academic Semester System. The Department Head and I decided to make this program an elective, it's an approved Tech Elec and a recommend Tech Elec for a couple of our tracks in ISE. We decided to make it "exclusive," meaning we wanted to have it have a positive brand, an "elite" image, something that would make candidates standout and a line item on their resume that would be a differentiator. We have come to learn over the past 10 years that this program has also become a differentiator for them when they interview, often their three semester experience is the focus of much conversation.

Figure 6.11 was the real catalyst for the program we developed. The Head noticed a gap between the way employers wanted their new ISEs to "think, talk, and behave" and the way we experienced them in their junior and senior years. Our employers wanted us to start the professional master transformation for them. Many of our employers also recognized our graduates were smart but also felt that they lacked an understanding of how to integrate all that they had learned in a way to rapidly create value for their organizations. The balance of soft skills and hard skills wasn't well developed.

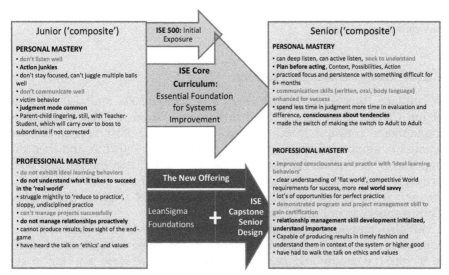

* A fact based yet personal representation of before and after.

FIGURE 6.11
The motivation for the ISE ILSS certification program at OSU.

Many ineffective habits existed that they then had to work on changing. So, in general, this list of "habits" on the left side of Figure 6.16 was what we experienced and the list of habits on the right side was what industry wanted. We designed our program to certify ISEs in ILSS but at the same time create a shift in both personal and professional mastery. That's probably the most important aspect of this program.

With that as background, let's turn to the Components of the Program as it currently exists.

6.2.3.3 Concept Design

The design of this new ISE Program Component at OSU in 2007 really, as mentioned, emanated from a desire on the part of both the Department Head and myself to create an experience (learning, developmental) for undergraduates (primary focus) that would better ready them for the first 3–5 years of their professional careers (Sink 2013, 2016; Sink and Higle 2009). We both felt that the core curriculum in ISE was adequate but that the students weren't as prepared to contribute, add value in the real world as they should be. We created a "composite profile" of students in their junior years and then an ideal profile of how we wanted them to "be" in their last semester. Note that the profile is really a list of what you'd generally see as being soft skills.

We intended to blend in personal and professional mastery with their ISE Core and this new three semester program in ILSS.

Let's now focus on the components of the new program and how we blended them together.

6.2.3.4 Components of the Program

We utilized Peter Senge's "Fifth Discipline" (Senge 2006) and also elements of Jim Collins' learnings in Built to Last and Good to Great to serve as foundational concepts in our design (Collins 2001, 2004). Simply put, we wanted to emphasize the importance of "discipline" in ILSS and then the importance of the balance between the "soft skills" and the ISE and ILSS principles and methods that are part and parcel to this type of training. We also utilize the notion of "full potential performance" (Figure 6.12) that Jim Collins speaks to in Built to Last.

We also emphasized to the candidates that this was 'training and education' not just education and introduced them to various versions of Bloom's taxonomy and kept emphasizing that we were working to get them to levels 3, 4, and 5 of the taxonomy—Apply, Analyze, Evaluate, and eventually Create.

This plays out in terms of curriculum design.

We utilize a "blended training model." Its components are as follows:

- Moresteam's BB, online, web-based core curriculum (with Minitab)
- OpEx ILSS and Minitab
- Deloitte Touche Enterprise Value Map
- ILSS Tool Book
- Learning to See

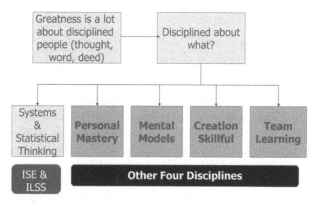

FIGURE 6.12
Blending Collins and Senge as the training framework.

- And then hundreds of support case studies and past project stage and gate decks to use as prototypes and/or examples. Final Tollgate Articles from iSixSigma and case study material from our own projects are also included.

Candidates complete the Moresteam Black Belt Foundation Course, 13 Chapters, DMAIC, approximately 140 h of study time prior to the training beginning (over winter or summer break). That enables us to be able to migrate up the Steps to Knowledge and Skill Taxonomy Ladder and get them into applying things right out of the starting gate. Our First of Five Labs, the Lean Lab, is the first Saturday of the training.

There are five "labs" in the 14-week BB Foundation Course:

1. A Lean Lab—we use what's called "Sticklebrick" (Prod Sim 2018) as our physical simulation. With a class of 28–34 we would have two supplier/production teams and then one small customer team. It's a very robust Lab that allows us to work with learnings from the all day (Saturday) lab for 2–3 weeks. The knowledge/skill components are as follows:

   1.1 Process capability analysis

   1.2 Measurement system analysis

   1.3 Measurement and analysis planning

   1.4 Flow, cells, kanbans, etc.

   1.5 Simulation, we have them use process playground as a tool to experience building and working with a simulation model of the process

   1.6 Team work

   1.7 The value of having process improvement specialists trained in DMAIC (in the third production cycle, we introduce students who are in the two semester capstone project portion of the program and they guide, facilitate a Kaizen-type event leading up to the third and final production run.

   1.8 They do DEFINE, MEASURE, and ANALYZE stage and gate presentations, so they are internalizing DMAIC as part of the lab.

2. A "SixSigma"-focused lab—we use what's called "catapult" as the physical simulation for this lab. This lab is run with three person teams. The knowledge/skill components are as follows:

   2.1 Process capability analysis

   2.2 Systems dynamics modeling

   2.3 Measurement systems analysis

2.4 Design of experiments (factor screening, n-factor, two-level DoE, use of Minitab and optimizer, etc.)

2.5 Importance of Standard Operating Constraints (SOCs), Standard Operating Procedures (SOPs), standard work, control of noise and other factors that create variation, etc.

2.6 The focus in this lab is more on MAI and C, so once again they are internalizing the DMAIC methodology.

3. A "Soft Skills," Change Leadership and Management Lab—we utilize an experiential set of exercises developed by Scott Sink, John Webb, and David Poirier called "boot camp." We focus on Senge's Other Four Disciplines: Personal Mastery, Mental Models, Team Development, and Creation Skillfulness (Senge 2006). The knowledge/skill components are as follows:

3.1 Power of habits

3.2 Attitude is a choice, intentionality, mind-set management

3.3 Professional modes of functioning, change management requirements

3.4 Speed of trust

3.5 Feedback process

3.6 The creation skillful process and how to employ it for process improvement projects

4. An Operational Analytics Lab—we utilize a variety of data sets many of which are actual data sets, disguised, from actual projects we have completed. This lab is done before, during and after weekly training sessions (that run 4 h) and spans 3–4 weeks of the 14 weeks. The knowledge/skill components in this lab are as follows:

4.1 Measurement System Analysis—give them data sets with weak operational definitions, poor organization, ambiguous column heads, etc., and sensitize them to the realities of data extracts in real world

4.2 Heavy work on the Foundational Data Role as well as the Analytics, Interpretation and Portrayal Role, lots of practice

4.3 Exploratory Data Analysis—practice with creative visualizations of various types of data, work on chart and graph skills (content as well as form and style)

4.4 Confirmatory Data Analysis—practice with staged control charts, hypothesis testing (picking right test), etc.

4.5 Reporting findings—creating mini decks that clearly and concisely report findings, answer the fundamental questions that were being addressed.

5. The final, fifth lab, is a "summative/capstone" lab that is intended to bring it all together, DMAIC wise. We utilize Moresteams SigmaBrew Simulation (online). The knowledge/skill components in this lab are as follows:

   5.1 How to prepare for and run/lead a Stage Gate Meeting

   5.2 Various types of meeting support materials that can be used to ensure a great meeting. The candidates practice: standard PowerPoint presentations; butcher block walls that display the Value Stream Map, Failure Modes, etc.; Prezi style presentations; elevator pitch style (no support material at all).

   5.3 This is the last pass through DMAIC so I'm always on the lookout for subtle key points that might not be understood about what happens in each stage of DMAIC.

That's a quick, outline version of the design of the training. For more detail, please feel free to contact Scott Sink (sink.22@osu.edu).

6.2.3.4.1 *Evaluation after 10 Years*

We have now run 525 ISE students through the BB Foundation Course since winter 2008. Over 200 have also taken the Capstone Certification 2 Semester Sequence for their GB certification. (Scott also trained and certified over 200 while at Global Life Sciences International). We do midsemester and end of semester course and instructor evaluations, Senior Exit Interviews, and monitor postgraduation feedback as best we can and our alums provide.

A general but representative summary of student feedback would be the following:

- Hardest course I've taken, best course I've taken
- Was worried I couldn't do it, once I got into it my confidence grew
- Not the path of least resistance but I'm more disciplined and skillful because of this
- It was all we talked about in my interviews, definitely set me apart resume wise and interview wise
- Helped me with my last Summer Internship or I know better understand what I was doing on my last internship
- Highly recommended by those that have taken it so I decided to try it

I would be remiss if I didn't also say that there is a segment of students, I'd say 10%–20% that think it's not a valuable experience, didn't like the instructor, and didn't like the material, don't think this is what ISE is about, don't want to do this type of work.

95%+ of the candidates taking the three course sequence (ISE 5810, 5811, 5812) had a job they wanted before they graduated. That's contrasted with approximately 78% for ISE and 70% for the overall College of Engineering. So, the initial success rate appears to be a bottom line measure of success.

6.2.4 Jabil Inc.'s Lean Six Sigma Program Overview

6.2.4.1 Jabil Introduction

Jabil (NYSE: JBL) is a product solutions company providing comprehensive design, manufacturing, supply chain, and product management services. Founded in 1966, Jabil's 180,000+ employees operate from over 100 facilities in 29 countries and had revenues of $19B in FY2017. Jabil delivers innovative, integrated, and tailored solutions to customers across a broad range of industries. Jabil's mission states that the company "is committed to becoming the most technologically advanced manufacturing solutions provider, who cares about people and the planet." To achieve these goals, Jabil has fully embraced Lean Six Sigma.

6.2.4.2 Jabil Lean History

In 2005, Jabil began the first corporate-wide Lean initiative. The focus of this initial effort was based on the implementation of Lean tools and was led by the Quality team. After a few years, the effort failed to achieve the desired results and eventually faded away. The lessons learned from this experience are basically twofold. First, a Lean effort needs to be transformative in more than Tools in order to sustain itself, so the next effort needed to look toward cultural transformation and not just application of Tools. Second, the Quality team had too many other priorities that took their focus away from developing and nurturing a Lean culture. The Quality group had many sound reasons that took the focus away from Lean—ISO audit, customer visit, etc., so the next effort needed a team solely focused on Lean transformation.

In late 2009, Jabil under our current CEO—Mark Mondello, but at the time our COO, wanted to restart the Lean program. Taking into consideration the lessons learned from the effort in 2005, a Lean Director position was created reporting to the Senior Vice President for Global Operations. The first request from the COO was that all Operations Managers in Jabil would become certified as a BB. The impetus for this early request was to provide the leader of each operation the credentials to become better leaders. In addition, there was a marketing aspect to this request to show our customers that our leaders are of the highest caliber. Therefore in early 2010, we began our first certification program focused on Six Sigma. In order to set a credible standard, Jabil had all students sit for the American Society for Quality Black Belt—Certified Six Sigma Black Belt (CSSBB) certification.

At the same time as developing our Operations Managers in Six Sigma, we had a decision to make that I believe all companies that embark on a continuous improvement path need to make. Did we want to become a Lean company or a Six Sigma company? We decided that our direction would be to become a Lean company and Six Sigma would be used to support the Lean efforts. For us, this is a very important distinction and impacted our education programs. Upon making this key decision, we taught the Six Sigma BB course as a transformation course and not solely Six Sigma. This distinction helped us lay the foundation for our overall objective to transform the company into one focused on Lean. After several years of Lean deployment, this model has proven to add value toward our objective to change the culture whereby everyone can participate. The latest analysis indicates that approximately 97% of all projects are Lean based whereas the balance requires the rigor of Six Sigma.

To be a Lean company, we wanted everyone to be engaged and involved regardless of their position. Therefore, we developed a Lean Certification for every level of the organization. These certifications are described in Figure 6.13.

6.2.4.3 Champions Training

This 1-day course is designed to teach leaders how to be effective in deploying and supporting Lean Six Sigma within their organizations, to foster alignment

FIGURE 6.13
Jabil's Lean Six Sigma program.

within the site Leadership Team, and to gain consensus on the commitment of embracing Lean and driving Cultural Transformation from the top down. We use Champion Training as the first step in a site's lean journey.

6.2.4.4 Lean Leadership Academy

This is a boot camp-styled workshop designed to introduce managers and leaders to all facets of Jabil's Lean education. It is a blended-learning approach that combines simultaneous in-class lectures as well as practical learning centered around team-based simulations, most intensive of which is a live simulation of a manufacturing floor complete with all players that one would find in any manufacturing value stream. This exercise enables the student to understand the use (or misuse) of the Lean tools as well as what needs to be in place before Lean tools can be used effectively.

6.2.4.5 Shop Floor Engagement Program

Jabil believes in leveraging collective efforts in maintaining the spirit of continuous improvement at the grassroots level in the organization, i.e., including the direct labor in making lean improvements. Keeping this in mind, the Shop Floor series was designed to equip operators with the essentials on lean and on basic wastes. The training was inspired by the Training within Industry approach and encourages operators to make small incremental improvements in their workplace while at the same time maintaining effective communication with their superiors to complete the feedback loop. The main focus is on spreading inclusiveness and engagement.

6.2.4.6 Bronze Level Training

The Bronze certification goal is to provide all the information needed to understand the basic concepts of Lean manufacturing and to develop effective practical problem solvers using the scientific method. The next training program up the "ladder" is aimed at the full-time employees from all departments. The Bronze training provides a more thorough training on the PDCA-based approach to problem solving. This includes introduction to lean concepts, tools for root-cause analysis, and different problem-solving methodologies. The concepts are reinforced through a set number of projects submitted toward the certification. In essence, Bronze training encourages practical problem solving no matter which work group the employee is based in and then opening the pathway for higher more intensive certifications.

6.2.4.7 Silver Level Training

Silver training is highly tactical in nature and is based on Jabil's Value Stream Transformation Model. Silver is designed to equip and train managers who are closely involved at the Gemba, to effect more aggressive performance gains and more challenging continuous improvement projects through kaizen events. Silver training teaches Jabil's value stream transformation system using the model of "teach the principle/concept/tool—then apply the principle/concept/tool." Silver encourages participants to not only work as part of a team to achieve a target but also helps them lead a value stream transformation kaizen event. Silver is targeted for highly selected student that is deemed to have high potential growth.

6.2.4.8 BB Certification

Six Sigma BB training provides an industry-leading approach that can tackle the complex problems and deliver breakthrough performance. Six Sigma is a highly specialized training involving the use of quantitative tools for process improvement. Most of the tools taught as part of the BB certification are data-driven tools and are heavily based on statistical concepts. Six Sigma is a disciplined, data-driven approach and methodology for eliminating defects in any process—from manufacturing to transactional and from product to service. It focuses on process improvement and variation reduction through application of Six Sigma improvement projects. In Jabil, Six Sigma BBs are developed through continuous Explain–Practice–Apply–Review learning cycle on in-depth knowledge of Six Sigma methodologies and tools. The development program consists of four 5-day training sessions over a 4-month period including one application project for each candidate.

Building a Lean Six Sigma Certification curriculum brought about an excitement throughout the company. People found the certification program challenging, fun, and meaningful to their workplace. However, there were some drawbacks to the certification program. Specifically, how does the certification fit into the larger Lean program?

6.2.4.9 Jabil Lean Implementation Plan

In the graphic (Figure 6.14), we devised a PDCA diagram to show how all the various components of a Lean program fit together. Specifically, Lean Education is found in the "Do" phase, but it is noteworthy to show where it fits in the larger framework.

6.2.4.10 Plan

The Plan phase starts with our strategy deployment alignment starting from our CEO to the Divisions and Corporate Functions and then to the Plants and all the way to the Shop Floor or Gemba. The cascading diagram (Figure 6.15) illustrates the catchball process from the CEO to the Gemba. In each location, we insure our alignment through a monthly Lean Council meeting where progress is reviewed, goals set, and resources allocated. On a regular basis,

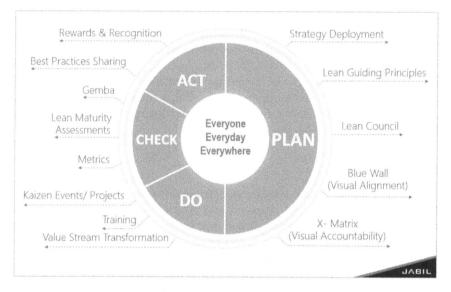

FIGURE 6.14
Jabil's lean implementation model based on the PDCA cycle.

the progress is reviewed at each level at a visual alignment board called a "Blue Wall." The Blue Wall is basically a blueprint for deployment and alignment. The Blue Wall varies depending on level of the organization, but the principles are the same. Specific actions are recorded and followed through an X-matrix and are the source of review and alignment.

Finally in the Plan phase, we seek to follow the company's Lean Guiding Principles largely based on the Shingo model. These Guiding Principles are as follows:

Alignment
- Create constancy of purpose.
- Align strategies, systems, and goals.

Cultural Enablers
- Lead with humility.
  - Recognize your own weaknesses to improve them.
- Promote team work.
  - Involve others in collaborative problem solving and decision-making.
- Develop people.
  - Challenge yourself and your team to continuously improve knowledge, skills, and abilities.

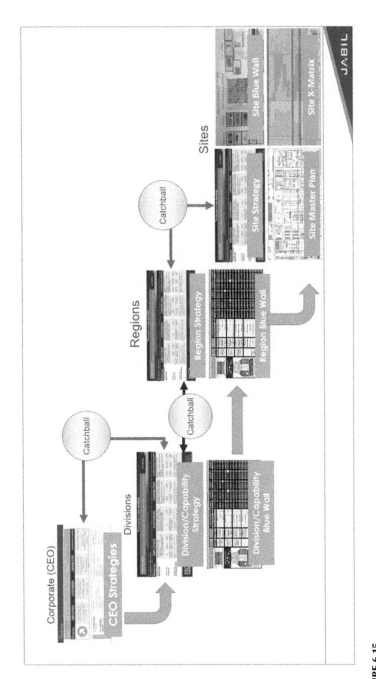

FIGURE 6.15
Jabil's "catchball" type of cascading and aligning targets.

Continuous Improvement
- Observe problems first hand.
  - Go to the Gemba to see and understand problems.
- Embrace scientific problem solving.
  - Solve problems with data and not opinions. Treat all employees as potential problem solvers.
- Relentlessly eliminate waste.

Build in Quality
- Make problems visible.
- Never pass a defect onto to the next process.

Velocity
- Focus on value streams.
  - Break silos. Understand your suppliers, inputs, processes, outputs, and customers.
- Eliminate anything that stops the flow of value creation.
- Create a pull system.
  - Produce only what is needed, when it is needed, in the right amount.

Results
- Create value for our customers.

6.2.4.11 Do

In the Do phase, we focus on education and certification to support the Kaizen events and to enhance the Value Stream Transformation. Therefore, Jabil's education and certification programs are intended to raise the deployment to that higher level of sustained performance. The blend of educational offerings along with the focus on Kaizen events drives toward that learning organization that we seek.

In the diagram (Figure 6.16), the focus at the Gemba is illustrated between "Running the Business" and "Transforming the Business." In the Running the Business section, we primarily focus our efforts on Lean Bronze and Shop Floor education in order to raise that level of understanding toward a high level of daily execution, whereas the people that are focused on "Transforming the Business" require more intensive education and coaching. This is where Lean Silver and BB programs add the most value.

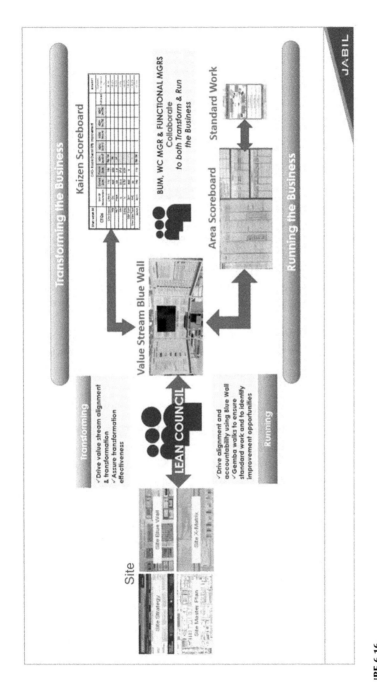

FIGURE 6.16
Jabil's standard visual management tools.

Date: December 2017	Overall Efficiency Improvements							Customer: XYZ		Last Update:	
Kaizen Title:	Value Stream Transformation Kaizen							Line/Area:		1/15/2018	
CTQs	Unit Of Measurement	Current State	Future State	KE #1 Sept 25-27	KE #2 Oct 09-13	KE #3 Nov 06-10	KE #4 Dec 11-14	KE #5 Jan 15-18		% Improved YTD	
MLT	Days	3.60	1.80	3.22	3.22	2.63	2.25	Upcoming		-37.50%	
Line Output	Units/h	110	150	132	132	145	155	Upcoming		40.91%	
WIP	Qty	422	209	330	290	220	220	Upcoming		-47.87%	
FPY	%	93.0	95.0	93.0	93.3	94.5	94.5	Upcoming		1.61%	
Scrap	%	0.53	0.30	0.50	0.50	0.42	0.42	Upcoming		-20.75%	
Crew Size	#HC	39	45	39	39	41	45	Upcoming		15.38%	
PPH	Parts per hour/HC	0.092	0.040	0.083	0.083	0.064	0.050	Upcoming		-45.83%	
Space	Sq-m	140	110	130	130	125	125	Upcoming		-10.71%	

JABIL

FIGURE 6.17
Jabil's value stream target sheet.

6.2.4.12 Check

The Check phase measures our success in removing waste and improving flow. We evaluate if the Kaizen events are helping each area to achieve the objectives set out every 6 months. In the chart (Figure 6.17), we show an example of a Target Sheet. After the completion of the Future State Value Stream Map, the team sets big transformational objectives that they believe can be achieved over the next 6 months and schedules Kaizen events over that period of time.

Additionally, Jabil's long-term sustainability is found in the Lean Maturity Assessments, which are intended to focus on the management adoption and leadership skills.

6.2.4.13 Act

Finally, in the Act phase, Jabil has a very strong Best Practice Sharing platform whereby great ideas, methods, and tools are uploaded to a company-wide intranet called ePromote. In ePromote, the Kaizen team is able to upload, search, and view Best Practices from throughout the company. This has reaped great benefits when a team in China is able to study a solution that came from Brazil.

Jabil is celebrating its 10th year conducting a company reward and recognition program called "Deliver Best Practices." Deliver Best Practices Competition is a year-long, global continuous improvement challenge to recognize, support, and promote continuous process advancements at Jabil. The competition showcases the best ideas employees implemented throughout the year using the Lean methodology to make Jabil more efficient, productive, and profitable.

The competition was created to engage employees; provide opportunities to share best practices and build bonds; and to generate excitement, awareness, and involvement. Because Jabil's priorities are the guiding principles that drive our culture of continuous improvement, this year's competition reflects upon our priorities of employees, customers, capabilities, and growth and the importance of maintaining a safety-first culture.

Projects can be submitted to one of the four categories which align with Jabil's priorities to help improve the company overall:

- Beyond the factory
- Employees
- Operations
- Social and Environmental

The finals are held at our corporate headquarters in St. Petersburg, Florida, every October, which is a huge celebration.

6.3 Summary

This chapter offers tested and integrated Lean Six Sigma approach from different industries and different sized companies as well as in the academic programs. If you are a leader in the Lean Six Sigma community, it offers some reference for you to deploy or improve your program. If you are an ISE student, it offers the motivation for you to understand the value of your training in this field during your UG or graduate study. If you are a BB or GB, it helps improve your understanding the thought process and your critical role in your journey toward operational excellence/continuous improvement. If you are a college or K-12 teacher, it gives you a practitioner's view how various Lean Six Sigma Programs are deployed and practiced in the industry and academia.

The authors all keep in mind that no programs are set in stone and must evolve and improve overtime. In this regard, we welcome feedback and exchange of ideas with each respective author after you have read this chapter. We also wish every reader an enjoyable journey in your OpEx/CI career and experience.

References

Collins, J. 2001. *Good to Great*. New York: HarperCollins.
Collins, J. 2004. *Built to Last*. New York: HarperCollins.
Feng, J. 2016. "Deploy lean 6-Sigma programs to organizations—Small and large." Invited speech to *The IMPACT Manufacturing Summit*. October, Chicago, IL.

Feng, J. 2017a. "Who said nothing is free?" Invited speech to *the American Manufacturing Strategy Summit*. October, San Diego, CA (jack.feng@cvgrp.com).

Feng, J. 2017b. Transforming a medium-sized company to better engage with both customers and suppliers. A phone interview published in February at *the Executive Platforms Thought Leaders Series—Lean Manufacturing Champions*. http://epthoughtleaders.com/?s=jack+feng.

George, M. L. 2002. *Lean Six Sigma*. New York: McGraw Hill.

Harry, M. and R. Schroeder. 2000. *Six Sigma*. New York: Currency.

Montgomery, D. C. 2013. *Introduction to Statistical Quality Control*, 7th edition. New York: Wiley.

Moore, G. A. 2001. *Cross the Chasm*. New York: Harper Business.

Prosci. 2000. https://www.prosci.com/.

Prod Sim. 2018. http://www.pro-sim.co.uk/.

Senge, P. 2006. *The Fifth Discipline*. New York: Doubleday.

Sink, D. S. 2013. "Fully minted industrial engineers: Ohio State's LeanSigma certification program puts graduates on the road to success." *Industrial Engineering*, December.

Sink, D. S. 2016. "Designing, developing, deploying performance improvement programs in the context of corporate transformations." *Moresteam Master Black Belt Working Lunch Presentation*, 23 September (copies available upon request, sink.22@osu.edu).

Sink, D. S. and J. L. Higle. 2009. "FUSION in the classroom—integrating lean sigma certification into an IE program." *Industrial Engineering*, August.

Snee, R. D. and R. W. Hoerl. 2003. *Leading Six Sigma*. Upper Saddle River, NJ: Prentice Hall.

7

How to Develop and Sustain a Lean Organization through the Use of Collective System Design

David S. Cochran and Joseph Smith
Purdue University Fort Wayne

Richard Sereno, Wendell Aldrich, and Aaron Highley
OJI Intertech, Inc.

CONTENTS

7.1 Introduction: Collective System Design Overview

Collective System Design (CSD) is a 12-step design process that provides organizations a roadmap on how they should function to meet customer and organizational objectives. This understanding is accomplished through a systems approach using lean principles as physical solutions (PSs) for meeting the functional requirements (FRs) of an organization. FRs define "what" the system must achieve, and PSs are chosen as the means to achieve those requirements. This system design language of the CSD methodology provides a way to communicate the requirements of the system. The effectiveness of the FRs and PSs are continuously being evaluated through the

Plan–Do–Check–Act (PDCA) learning loop for continuous improvement (Cochran et al. 2017). As outlined in the CSD 12-step design process, this transformation must start with Senior Leadership seeing a need and making a conscious choice to change to create value for the customer, organization, and society. The critical part is for Senior Leadership to see this need for change before it is a last attempt effort to save the organization.

To facilitate the understanding of the CSD methodology, the concepts will be explained in conjunction with a small manufacturer's recollection of their Lean implementation. The troubles faced throughout the implementation process are noted by the manufacturer. The CSD methodology is then discussed in detail along with the augmentation it could have provided to the manufacturer during their system redesign process had it been understood at that time.

7.2 The Lean Transformation from a Manufacturer's Perspective

As an example, a small manufacturer located in Northeast Indiana was facing an arithmetic problem. As simple as $2 - 3 = -1$, or if you sell a product for \$2 and your cost is \$3, you are left with a loss of \$1. The issue for the organization was clear, agreed upon collectively, and focused on the organization's cost structure. The costs of goods sold were too high, and the sales price of those goods was too low to cover operating expenses. To compound the simple arithmetic problem, the organization did not have control over large portions of their cost of goods sold, i.e., material, health care, and energy costs. These uncontrolled costs were increasing year over year.

Faced with no obvious or easy ways to reduce and/or even contain costs, the company's demise seemed certain. Reflecting on the current situation, the organization's management team clearly recognized the risks and challenges facing the organization. What was not clear were the many opportunities facing the organization. After a careful review of the financial statement, a few opportunities became apparent:

1. Scrap was too high.
2. The organization was carrying excessively high inventory levels and carrying costs.
3. Associate turnover was high.
4. The manufacturing operation was set up on mass production principles.
5. The organizational culture was top down, and the associates had little or no input into their work content or environment.

Based on the opportunities identified, the management team determined that the best way forward was to replace the mass production system with a manufacturing system designed to become lean. The leadership team agreed that if the organization was going to be successful, all levels and all associates would need to be involved and engaged in the system redesign process.

Moreover, the company was serving the automotive industry, where zero defects, 100% on-time delivery, and contractual year-over-year price downs were expectations, not targets. The management team understood that the lean transformation had to take place while it continued to flawlessly execute on its customer demand.

Despite the identified problems and potential opportunities, the leaders of the organization lacked the knowledge to implement the company's new manufacturing system design to become lean. To overcome this roadblock, the president of the organization recruited a plant manager who had 10 years of experience in the Toyota Production System as Senior Manager for Operations at a Tier 1 supplier to Toyota.

The organization was your typical mass production facility driven by traditional management accounting to maximize machine hours and utilization; the production schedules were disconnected from the actual customer demand (Cochran et al. 2014). The first lean system implemented by the new Senior Manager tied the customer demand to the production schedules using kanban and level planning in an attempt to stop overproduction, the worst form of waste (Ohno 1988). As expressed through the CSD methodology, the approach began with the implementation of level loading as the PS to meet the FR for producing the customer-consumed quantity per time interval.

A pull system was implemented as a PS on how to meet the FR of ensuring material availability even though fallout exists. The organization started the implementation by understanding the customer demand through the monthly demand forecasting from their customer. Once the monthly forecasted demand was understood, the demand information was leveled so that the daily customer demand requirement could be understood by everyone in manufacturing; the average pace of customer demand, called the takt time was established for each manufacturing line. Takt time (time/unit) is given by the following equation:

$$T = T_a/D \tag{7.1}$$

where T is the takt time, T_a is the net daily time available to work, and D is average daily the demand (customer demand).

The physical kanban quantity requirements were based on the average Overall Equipment Effectiveness (OEE) of the line (e.g., if a line is capable of producing 10 kanbans per hour at 100% OEE but has historically operated at 80% OEE, only 8 kanbans would be scheduled per hour. OEE is discussed in Section 7.4. When the customer pulled the product, shipping would remove the physical kanban and place it in a collection box to be redistributed to the manufacturing lines according to the level plans time interval replacement schedule based off of takt time. Figure 7.1 provides a sample value stream map of a pull system.

In order to visually manage to see if the manufacturing lines were producing to takt time, level planning with leveling boxes, sometimes called Heijunka boxes, was implemented at each line with time indicators to show if the line was producing to takt time (Liker and Meier 2006). Heijunka boxes were used to level both demand and product mix. Also, a Plan For Every Part (PFEP) system was implemented to visually show the minimum and maximum quantities required for the material store areas. As a measurement of the standard work in process inventory, the storage area was audited to see if the system was operating correctly within the minimum and maximum levels. As a result, on-time delivery was being measured to monitor the FR for replacing products that the customer consumed.

Through the implementation of the PS of level planning and tying production schedules to customer demand, the system design exposed overproduction waste. To correct overproduction, operations needed to be idled to right size the inventory to the capability of the processes. The need to idle the operations in the short term created an adverse effect on the income from operations as reported through the cost accounting system (Cochran et al. 2014). Running fewer machine hours equated to less applied hours being applied to offset the manufacturing cost. However, those machine hours were used to produce unneeded products in the months prior to the new system design implementation. With the manufacturing cost not being covered by the additional machine hours, the income from operations was coming in under budget which created concerns about the effectiveness of implementing lean systems throughout the organization.

Even though it was strategically decided that the organization needed to implement lean systems and the President recruited outside leadership

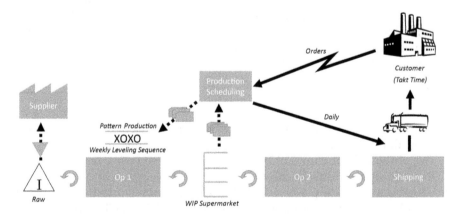

FIGURE 7.1
Value stream map of a pull system. (Adapted from Cochran et al. 2014.)

to lead the implementation, leadership did not fully understand the road that needed to be traveled to overcome the short-term impacts for greater results long term. It created questions on how implementing lean systems were going to benefit the organization by the Controller, the President, and in turn, the parent company in Japan.

Most organizations have seen the Toyota Production System house for the tools that are needed for a lean system, but it does not describe the tone that Senior Leadership needs to have to provide consistency of purpose throughout the entire organization. Without Senior Management collectively agreeing on the tone to establish the thinking, which in turn drives the creation of the process to satisfy customer needs, it becomes difficult to sustain lean long term. Some sources estimate that 90% of the companies that try to implement lean fail to sustain it after 3 years (Cochran and Kawada 2012; Womack 2011). The lack of sustainability could be due to viewing lean as a toolbox approach and not understanding that lean is a system design that requires not only support from Senior Leadership but a new mind-set and tone to implement.

7.3 Transformation versus Sustainability

The CSD Flame Model of a system illustrates that any system is made up of four elements that must work together as an integrated whole (see Figure 7.2). These elements are the tone, the thinking, the organization structure, and the actions and work. Figure 7.2 illustrates that system redesign starts with the tone and moves upward through the flame (Cochran 2007). CSD engages the people within an organization to gain collective agreement

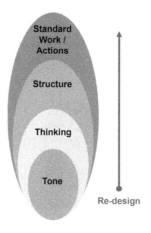

FIGURE 7.2
The Flame Model of Collective System Design.

about the objectives, called FRs, of the organization as part of the thinking layer of the Flame Model. The FRs of a system redesign define what an enterprise must accomplish to meet the needs of the customers internal and external to the manufacturing system. To sustain a system redesign implementation, everyone in an organization must agree on what the system seeks to accomplish and the means, called PSs, that are implemented to achieve the FRs (Johnson and Bröms 2011; Cochran 2013). Figure 7.3 illustrates that system redesign may be represented by three domains: the customer domain, the functional domain, and the physical domain.

The deep understanding of the Toyota Production System that the Senior Manager employed during system implementation required knowledge beyond the implementation of lean tools that fall within the physical domain. He knew that the FR of the system redesign was to produce the product that the customer demanded when the customer demanded it. Use of kanban cards, the leveling box, place for every part, and standard inventory were PSs that were implemented to achieve this FR.

The thinking layer of the Flame Model of system integration is expressed by a language for system design that expresses customer needs, FRs, and PSs. One PS is chosen and collectively agreed to by the design team to achieve one FR.

Tone is first. The people in an organization must have a voice and be heard. The implementation of lean tools as PSs without collective agreement by the people involved in the system redesign or new system design effort is typically not sustained.

Second, the thinking about the system redesign must be understood by everyone. The FRs of the system design must be defined based on what is most effective in meeting customer needs, NOT on optimization of financial targets. This idea is the fundamental difference in the Toyota Production System and Mass Production. Mass Production seeks to optimize financial targets driven by management accounting that optimizes parts of the whole system as independent islands of cost, whereas the Toyota Production System seeks to meet the needs of its internal and external customers with the least waste. The consequence of the system's implementation is to become lean. The term lean was first coined by John Krafcik to express

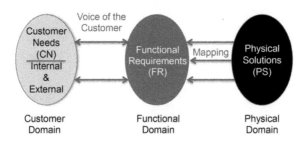

FIGURE 7.3
Three design domains. (Adapted from Suh 1990.)

the consequence and result of the Toyota Production System (Womack et al. 2007). Lean, therefore, is not what we implement as a system, but it is what we become as a result of meeting customer needs. The thinking layer of the flame uses a language to express and communicate a system design. An FR states what a system must achieve to meet an associated customer need.

The next layer of the Flame Model is structure. Structure pertains to organizational structure, plant layout, and the expression of material and information flow to meet takt time with a value stream map. Structure is implemented based on the PSs that are determined and collectively agree upon by team members as part of the thinking layer of the system redesign process. Structure or form of the organization follows function (Alexander et al. 2010).

Standard work and actions form the top layer of the Flame Model. Based on Dr. Deming's teachings, the Toyota Production System relies on a mathematical fact… that variation cannot be reduced, when the source(s) of variation is unknown (Deming 2000). To be able to improve a system, the work within a system must be made repeatable. Spear expressed the four rules regarding the DNA of the Toyota Production System (Spear and Bowen 1999). Rules 1 (all work must be specified according to content, sequence, timing, and outcome) and 4 (process improvement must be done based on the scientific method, under the guidance of a coach and by those doing the work closest to the problem) set the stage for why Standard Work and the PDCA process for improvement are necessary (Krebs 2008). The ultimate goal of establishing standard work is to treat work and actions as an experiment or as a type of experimental design to improve the PSs, which is the work

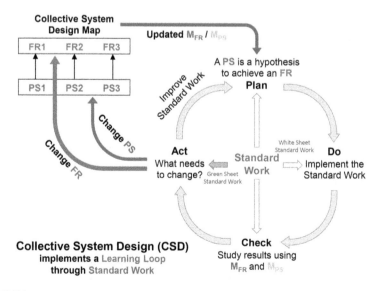

FIGURE 7.4
CSD map and PDCA relationship for meeting system FRs (Cochran et al. 2014).

that is proposed to achieve system redesign FRs (see Figure 7.4). When a failure or problem occurs in the implementation of a PS to achieve a FR, the team has three choices: (i) improve the standard work that implements a PS, (ii) change the PS and establish new standard work, or (iii) change the FR when the statement of customer need has changed.

7.4 The Difficulties of a Transformation from a Manufacturer's Perspective

Early on, standard work was primarily viewed by the company's leadership team as a means to conform to the International Standard Organization's (ISO) requirements. With regard to ISO, each element of the organization was audited to see if it had a procedure and if that procedure was being followed. This practice led to vague, text-heavy documents that were hidden away in electronic and paper folders that were only ever used during audits.

When the organization started to switch to a lean operating model, the leadership team started applying industrial engineering and training within industry (TWI) principles to standard work (Liker and Meier 2006). Processes were designed and balanced to takt time, the rate of customer demand, waste was eliminated, and employees were trained to the new procedures. Everything looked great on paper, but the reality on the shop floor wasn't much different than under the ISO model. A couple of key ingredients were missing: layered audits and employee involvement.

The organization was extremely flat. On the shop floor each cell would have a lead with very limited responsibility, meaning each of the 50 or so employees in each value stream reported directly to the Value Stream Manager, who was also responsible for everything from customer returns to supplier issues. This meant the manager had very little time to make sure a 5S program was being followed or to help an employee get a new wrench. The proposed solution, inspired by Toyota, was to introduce a new level of leadership, the Team Leader, to the organizational structure.

As with the level loading, adding a new position did not make any sense under the traditional management thinking and methodologies. Accounting was concerned with how adding a new top-tier hourly position would impact variable costs, sales per employee, and monthly payroll. Sales worried that the position would not be covered in the overheads used to calculate part price on quotes, so any business they acquired was going to look less profitable. Senior management had a hard time seeing the benefit of adding a person who would not be touching product and made note that the organization had functioned fine without this position for years. Everyone was focused on the trailing indicators tied to financial measures. Leading indicators, like how well employees were following standard work, were completely ignored. If the need for a new position had been expressed through the use of CSD, the FR to respond rapidly to production disruptions would have driven the need for a Team Leader position. Understanding the system as a whole and the requirements of that system, the need for a Team Leader position would have been better communicated.

Another tool that was implemented was OEE. OEE quantifies how well a manufacturing system performs during the periods in which it is scheduled to run. This metric is determined by taking the product of a lines uptime (availability), quality, and performance (actual performance rate as a percentage of the theoretical performance rate) given by the following equation:

$$OEE\% = (Availability\%) * (Quality\%) * (Performance\%) \qquad (7.2)$$

This number gives an idea regarding the theoretical capacity of a line given its current scheduled hours. For example, if a line has an OEE of 65%, it could theoretically produce 35% more product during the same time frame (Williamson 2004). OEE was employed with the help of electronic monitoring boards attached to each line that gave real-time information on how the line was performing. OEE was one lean tool that made sense to traditional management thinking. OEE's origin is from the 1960s and mass production thinking (Williamson 2004).

OEE was a technology-based solution that the company installed that measurably drives up performance and which has been traditionally used to beat people over the head with if they step out of line. It was also a natural fit for an incentive system. This approach led to a rapid $40,000 in savings, but there were hidden long-term consequences. OEE naturally aligned with point-solution applied thinking, a higher OEE means higher utilization and more applied and efficiency gains. So, management became focused on driving OEE higher, but with the OEE incentive system installed, the production operators wanted to keep their line's OEE just above the target, so as to get the most gain out of the least amount of work, thus, creating conflict between the two groups. Notice who has been completely left out of this discussion, the customer.

What would have provided the organization a more consistent approach, so everybody understood the impact of implementing lean systems?

7.5 Applying CSD Methodology to Sustain System Design Change

The thoughts and actions associated with the belief that an organization can implement lean stands in stark opposition to the fundamental goal of achieving the FRs that result in a system to become lean. When lean is defined as the result of a system, designed to meet the needs of the customer(s), then the focus turns to sustaining a transformation by understanding the system as a whole. Viewing lean as a tool box of solutions results in applying point solutions to optimize parts of the system. Without an understanding of the customer needs and FRs of the system to achieve customer needs, the result of implementing the various lean tools can have debilitating consequences.

A Collective System Design Map is developed as part of the 12 steps of CSD. This map defines the FRs of a system and purposes PSs to meet each of the FRs. The interactions among the PSs and FRs are assessed to ensure the path dependency of implementation is understood (Suh 2005). Once the Collective System Design Map is complete, the leadership of an enterprise will understand the implementation sequence for the transformation. This implementation sequence allows the chosen PSs to be implemented in a predetermined sequence that streamlines implementation.

What is the result of not following or understanding the implementation sequence of a transformation? Consider an example of implementing single-piece flow. Upon entering a mass production facility, a team notices that batch production is the method being used. With the goal of "implementing lean," the team decides to restructure the operation to follow single-piece flow. After a brief time, production ceases as a consequence of poor quality parts. The operators cannot continue to work if the previous operation hands them a defective part. With this discovery, the team decides that quality must be improved. As the team begins to look for the source of the defects, they find that defects are occurring in almost every operation of the manufacturing system. The team decides that in order to find the source of variation, they must ensure the work of the operators is consistent (i.e., standard work).

In this example, understanding that standard work drives the ability to recognize problems, and recognizing and solving problems are the key to successfully implementing single-piece flow, the team could have avoided much of the difficulties experienced. In the above-mentioned example, the top-level manager, who was not familiar with lean tools, doubted the capabilities of the Senior Manager's lean ideas after the initial implementation of single-piece flow because production output decreased. An understanding of and collective agreement regarding the FRs of the manufacturing system to meet customer needs, through the use of CSD, would have provided the top-level manager with the knowledge of the implementation sequence of the system design PSs and the tone required to sustain the system design implementation.

The Manufacturing System Design Decomposition (MSDD) map was developed by Dr. David Cochran and his research team at MIT to provide a Collective System Design Map defining the FRs of any manufacturing system (Cochran et al. 2001/2002). An overview of the MSDD presented

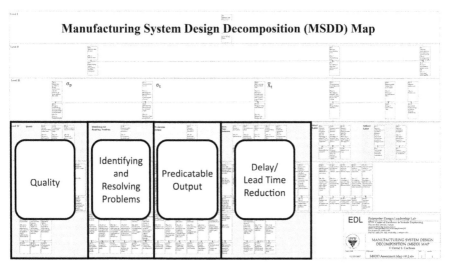

FIGURE 7.5
MSDD Map: A Collective System Design Map for Manufacturing.

in Figure 7.5 shows at a high level the implementation sequence (left to right) of a system design implementation. On the far left of the MSDD is the Quality branch (which includes implementing standard work) followed by the Identifying and Resolving Problems branch. The next branch is for ensuring a Predictable Output, and finally, the Delay/Lead Time Reduction branch which includes single-piece flow. Thus, the map establishes a starting point for improvement that relies on standard work.

The MSDD map guides the implementation of a manufacturing system design by outlining the implementation sequence of the PSs of the design; some of which are lean tools (Cochran and Swartz 2016). The MSDD map not only guides the implementation sequence but it systematically removes the seven wastes by defining a system that meets the needs of the customer. The FRs and PSs that tie to the seven wastes are identified in Figure 7.6. The mapping of the seven wastes to the MSDD map is presented in Figure 7.6 as well. Ohno said that "the basis of the Toyota Production System is the absolute elimination of waste" (Ohno 1988). Toyota originally identified the seven major types of non-value-adding activities within an enterprise, but an eighth waste of unused employee creativity was presented in The Toyota Way Fieldbook (Liker and Meier 2006). Although the MSDD map does not address the waste of unused employee creativity, the flame model and the CSD 12 steps, presented below, allocate an entire step (Step 3) to facilitating enterprise-wide participation through the Establishment of Tone and Values.

Waste	Description	Functional Requirement	Physical Solution
1. Over Production	Produce sooner, faster or in greater quantity than customer demand	FR-T22 Pace manufacturing system to takt time	PS-T22 System (cell) designed to meet takt time
2. Inventory	Raw material or work in progress not having value added to it	FR-T23 Pace part arrivals to takt time	PS-T23 Information system designed to meet takt time
3. Waiting	People or parts waiting on a work cycle to be completed	FR-D1 Eliminate operators waiting on machines	PS-D1 Human-Machine separation
4. Motion	Unnecessary motion of people or movement of parts	FR-D2 Eliminate wasted motion of operators	PS-D2 Design of workstations/work-loops to minimize unnecessary motion
5. Transportation	Unnecessary movement of people and parts between stations	(People) FR-D21 Minimize wasted motion of operators between stations (Parts) FR-T1 Reduce lot delay	PS-D21 Machines/stations configured to reduce walking distance PS-T1 Single-piece flow cell design
6. Rework	Not making the part correctly the first time: requiring the need for correction	FR-R1 Respond rapidly to production disruptions	PS-R1 Procedure for detection & response to production disruptions
7. Over Processing	Processing beyond the customer defined standard	FR111 Manufacture products to design specifications	PS111 Design of manufacturing processes with minimal variation from the target

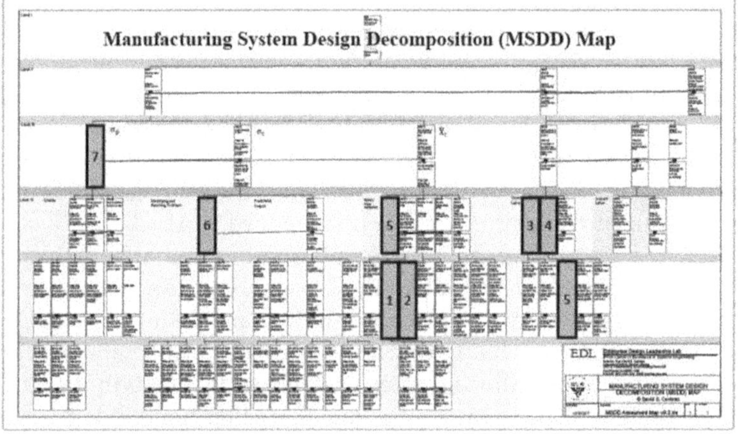

FIGURE 7.6
The seven wastes identified in the MSDD map.

7.6 The CSD Flame Model and the 12 Steps

The CSD Flame Model presented in Figure 7.7 represents the elements that coexist simultaneously within a system that includes the thoughts and actions of people. Diagnosis is analogous to entering the flame, while the system redesign to become lean is the result of emerging from the inner part of the

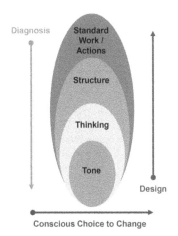

FIGURE 7.7
CSD Flame Model for system design.

flame (Cochran et al. 2016). The entirety of the diagnostic process is to understand the root cause of the problem. Shigeo Shingo articulated understanding the root cause by asking "why" five times (Shingo 1989). The diagnosis via the flame model begins in the outer layer by recognizing the ineffectiveness and waste associated within the work layer. Then, the next layer begins to understand the structural problems, such as misaligned metrics or policies, associated with the problems identified in the work layer. Continuing towards the center of the flame, the thinking layer seeks to understand the design thinking that leads to the problematic structure. Finally, the hottest part of the flame (the center) works to discover the tone of management that drives that thinking.

Transitioning from the diagnosis phase to the design phase begins the 12 steps of CSD by the Senior Leadership making a conscious choice to change (Step 1). The flame model, acting as the heart of the CSD approach, defines the design phase as one moves from the hottest part, the tone, of the flame toward the outermost part, the work and actions. The tone refers to setting the right mindset, attitude, and perspective. A deep respect for everyone and everything involved (i.e., customers, suppliers, employees, the environment) should fall within the tone defined by senior management. This tone drives the thinking phase of the design. Within the thinking phase, the logical design is developed through a system design map driven by understanding the needs of the customer (Cochran et al. 2014). From the thinking layer, the structure of the organization will be realized. The structure refers to the physical system that should change. From here, the work/actions are defined. These include the processes and standard work that are required to ensure the change is effective.

The structure of the flame defines the various layers within the CSD methodology. Presented in Table 7.1, the 12 steps are identified and the important

TABLE 7.1

The 12 Steps of CSD and the Associated Questions

Step	Descriptions	Questions
1	Senior leadership makes a conscious choice to change	• Why would we make the change? • Are we capable of achieving something greater? • Is continuous improvement important to us?
2	Define stakeholders and system boundary/value stream(s)	• Who will be affected by the change? • Who should be involved in the process? • What/who can and cannot be controlled? • What information is passed across the system boundary? • What risks exist within the interfaces?
3	Establish tone and values	• What attitude is required to get everyone to participate? • What attitude is required to facilitate collective agreement? • How do we convey to our people the tone we desire for our organization? • How do we problem solve together?
4	Identify customers and needs	• Who will purchase/use our product(s)? • What will this customer/user need the product to do?
5	Determine FRs	• How do we state the needs of our customers as FRs of the system? • What MUST the system achieve to satisfy the customer(s)?
6	Map the PSs to FRs	• What function MUST the system achieve? • Is the design uncoupled/partially coupled but not fully coupled? • Are the requirements of the leaf PSs sufficiently clear and implementable through standard work?
7	Define performance measures (FR_M and PS_M)	• How will we know if the FR is achieved (FR_M)? • How will we know if the PS is implemented correctly (PS_M)?
8	Define organization structure based on CSD map	• What team structure is needed to make the implementation? • What should the value stream look like? • Can we physically simulate the value stream?
9	Establish standard work by continuous improvement: PDCA	• Currently, what is the best practice for completing the work? • How will we implement a PS as standard work? • Does the current standard work achieve the FRs of the system design? • How can the standard work be improved?

(Continued)

TABLE 7.1 (*Continued*)

The 12 Steps of CSD and the Associated Questions

Step	Descriptions	Questions
10	Evaluate the cost of not achieving the FRs	• What is the result of not achieving each one of the FRs? • What is the cost benefit of achieving, unachieved FRs?
11	Prepare resource reallocation plan	• What restructuring of resources is required to make and sustain the system design transformation?
12	Feedback for sustainability and growth	• What was the result of the implementation? • What are the required continuous improvement efforts that are necessary to sustain and to improve the system?

questions that one should consider are provided to help understand the consequence of each step. Consider the questions associated with Step 3. These questions should provide a basis for understanding the tone and mindset necessary to sustain change. The tone will be unique to each company, and the process of understanding the tone is arguably just as important as practicing a tone that brings health to an organization. Facilitating collective agreement, and getting everyone to participate (i.e., operators who want to follow standard work) requires the humility exhibited by the Senior Leadership team. This realization once understood and conveyed to the enterprise, will provide the foundation for sustaining change.

7.7 Industry Reflection on the Use of CSD

The CSD Flame Model of a system provides an overview of how the small manufacturing company could have used CSD from the beginning of their lean journey to provide a working framework to bring consistency of purpose between the leadership being brought in from the outside to implement lean systems, President, Controller, and the parent company in Japan. The Senior Leadership within this company made a conscious choice to change (Step 1) and recognized what was and was not in their control (Step 2). Step 3 begins the crossover point of how this manufactures transformation may have differed through the application of CSD.

The CSD 12-step methodology would have walked the company's senior leadership through the need to collectively agree on the tone the organization needed (Step 3) to take to drive their thinking in understanding customer needs (Step 4) and the associated FRs (Step 5). Once the FRs were established, it would have walked them through how to implement the lean tools to address how they were going to meet their FRs (Step 6). Step 6 would have defined the relationship between the lean tools as PSs to meet the FRs of the manufacturing system and how the implementation sequence of the PSs can be path dependent. Finally, the effectiveness of the system would have been measured through FR and PS measurements FR_M and PS_M (Step 7).

In the case of implementing level loading, the relationships defined in the MSDD map would have shown a path to right sizing the inventory levels to meeting customer demand. The FR to reduce process delay is achieved through producing to takt time. The FR measure (FR_M) of reducing the process delay would have indicated excessive inventory through not producing to takt time. This indication would have shown a need to reduce machine hours, and the affected income from operations in the short term would have been better understood. The Senior Leadership could have used the CSD 12 steps as a road map for communicating how the system redesign to become lean would impact the organization short term as well as long term. Defining the structure of the organization (Step 8) would have detailed the need for the Team Leader position and reduced the confusion with Senior Leadership during implementation.

The PDCA cycle can be used to validate the right solutions and continuously improve them (Step 9) (Deming 2000). Once the plan is put in place for implementation, the measures can be checked for the FRs and PSs to see if the system design implementation is meeting expectations. If the FR measures are falling short of expectations, the necessary changes could be made. In the case of level loading, if the inventory levels were adjusted too low in the presence of too much variability in the process, the FR for on-time delivery would not have been met. Process capability could have been evaluated so that an adequate level of inventory could have been determined to accommodate for the presence of variation.

The MSDD map addresses right sizing the inventory to match process capability through the FR of ensuring parts are available to the material handlers. This FR is achieved through a marketplace of standard work in process between subsystems. The associated FR measure is based on the number of marketplace shortages. Therefore, the number of shortages within the marketplace of standard work in process would have driven the reduction in inventory to match the variation in the system.

After the implementation of standard work and the PDCA cycle, Senior Leadership would have better understood the financial benefit of achieving the system FRs within the MSDD map (Step 10). From here, preparing a resource reallocation plan from a perspective of meeting the system's FRs could have been created (Step 11). Finally, feedback based on the system design and details regarding the need for continuous improvement efforts would be relayed to Senior Leadership (Step 12).

7.8 Engagement in Standard Work through Tone

To provide a more detailed understanding of the tone required to get everyone to participate, the following example details the trouble experienced by a manufacturer seeking to implement standard work. After their third attempt, the standard work begins to take root. The details of the three attempts are presented below.

During the first attempt at implementing standard work, the engineers developed the Standard Operating Procedures (SOPs). The manufacturer recalls the delivery method of these SOPs as simply, "throwing them over the wall." As one may expect, this method did not result in the sustainable implementation of standard work. The actions and attitudes of the associates, as recalled by Senior Leadership, were similar to "I am not a robot" and "What do you need me for?"

During the second attempt at implementing the SOPs, Teams Leads were involved in the discussion regarding the SOPs. In the end, the SOP was still developed by the engineer with very little input from the Team Leads. The delivery method to the associates during this attempt was through the Team Leads. In addition, the Team Leads were responsible for auditing the workers to ensure that the SOPs were being followed. The manufacturer recalls that this attempt resulted in workers quitting. The team decided to pull back and take another look at the process of implementing SOPs.

During the third attempt, the manufacturer recalls working with everyone in a room except managers and engineers. The associates in the room were given the task to define and agree on the SOP for the various operations they were involved in. TWI began to be used at this time to help sustain standard work. After some time, the team came to an understanding that the documentation of the TWI and SOP belonged to the engineers and Team Leads, but the actual process of standard work needed to belong to the associates. Only after management began to view themselves as a supporting role, and the workers began playing a key role in developing the standard work, did the implementation of standard work take root within the company.

7.9 Conclusions

Lean is the result of a system design or system redesign and should not be the activity that we go do. The PSs to achieve the FRs that meet customer needs is the focus of the Collective System Design Process that results in becoming lean. Sustainability of a system design or redesign occurs because of clear communication of FRs and PSs, a tone by leaders in the organization that all abnormal conditions are opportunities for improvement and the emphasis that issues can only be identified and resolved when standard work is followed by everyone in the organization. The example provided in this Chapter describes how Collective System Design enhances lean from a comprehensive system design perspective that includes lean thinking, lean tools, and lean philosophy. The recap of the small manufacture's journey, in conjunction with the 12 Steps of Collective System Design, details a system design process to become lean.

References

Alexander, C., Ishikawa, S. and Silverstein, M. 2010. *A Pattern Language: Towns, Buildings, Construction*, New York: Oxford University Press.

Cochran, D. S. 2007. Systems approach to sustain lean organizations, *2007 SAE International World Congress*, Cobo Hall, Detroit, MI.

Cochran, D. S., 2013. Chapter 18, Sustaining the Lean Enterprise, In *Lean Engineering*, Black, J. T. and Phillips, Don T., Virtualbookworm.com Publishing.

Cochran, D. S., Aldrich, W. and Sereno, R. 2014. Enterprise engineering of lean accounting and value stream structure through collective system design, Abstract 55, *Institute of Industrial Engineers, Engineering Lean and Six Sigma Conference 2014*, Orlando, FL.

Cochran, D. S., Arinez, J., Duda, J. W., and Linck, J. 2001/2002. A decomposition approach for manufacturing system design. *SME Journal of Manufacturing Systems*, Vol. 20, No. 6. 371–389.

Cochran, D. S., Elahi, B. and Spurlock, T. 2017. System re-design of first tier automotive value stream, Abstract 3368, *Institute of Industrial and Systems Engineers (IISE) 2017 Annual Conference and Expo, Industrial and Systems Engineering Research Conference*, Pittsburgh, PA.

Cochran, D. S., Hendricks, S., Barnes, J. and Bi, Z. 2016. Extension of axiomatic design theory to implement manufacturing systems that are sustainable, *ASME Journal of Manufacturing Science and Engineering: Special Issue on Sustainable Manufacturing*, Vol. 138, No. 10. 1–10.

Cochran, D. S. and Kawada, M. 2012. Education approach in Japan for management and engineering of systems, *2012 ASEE Annual Conference and Exposition*, San Antonio, TX.

Cochran, D. S. and Swartz, J. 2016. Sustaining improvement through tone in collective system design, Abstract 1707, *Institute of Industrial Engineers 2016 Annual Conference and Expo, Industrial and Systems Engineering Research Conference*, Anaheim, CA.

Deming, W. E. 2000. *Out of the Crisis*, Cambridge: MIT Press.

Johnson, H. T. and Bröms, A. 2011. *Profit Beyond Measure: Extraordinary Results through Attention to Process and People*, London: Nicholas Brealey Publishing.

Krebs, D. 2008. The Mysterious 4 Rules. Lean Healthcare Exchange. www.leanhealthcareexchange.com/the-mysterious-4-rules/.

Liker, J. K. and Meier, D. 2006, *The Toyota Way Fieldbook: A Practical Guide for Implementing Toyotas 4Ps*, New York: McGraw-Hill.

Ohno, T., 1988, *Toyota Production System: Beyond Large Scale Production*, Cambridge, MA: Productivity Press.

Shingo, S., 1989, *Study of the Toyota Production System from an Industrial Engineering Viewpoint*, Cambridge, MA: Productivity Press

Spear, S. J. and Bowen, K. 1999. Decoding the DNA of the Toyota Production System. Harvard Business Review.

Suh, N. P. 1990. *The Principles of Design*, New York: Oxford University Press.

Suh, N. P. 2005. *Complexity: Theory and Applications*, New York: Oxford University Press.

Williamson, R. M. 2004. *Don't Be Misled by O.E.E*, PDF. Columbus: Strategic Work Systems, Inc.

Womack, J. 2011, Keynote Speech, Lean Accounting Summit, Orlando, FL.

Womack, J. P., Jones, D. T. and Roos, D. 2007. *The Machine that Changed the World*, London: Simon & Schuster.

8

Building and Managing the Bill of Process to Streamline the Enterprise—An Emerging Technology-Enabled Systems Approach

Dave Sly
Iowa State University

Carl Kirpes
Marathon Petroleum Corporation

CONTENTS

8.1 The New Age Industrial Engineer— Value in the 21st Century

Is industrial and systems engineering truly valuable? Do industrial and systems engineers (ISEs) create value for their organizations? Since the days of Frederick Taylor, ISEs have worked in organizations performing time studies, generating line balances, writing work instructions, completing ergonomics assessments, ensuring quality, and completing Process Failure Modes Effects Analysis (PFMEA). However, as organizations have grown more complex, often different ISEs are performing varying degrees of each of these functions for their company, and in academia, different ISEs are doing research and teaching specific segments of these topics. This siloed approach allows for deep knowledge and strength within a discipline but ignores the opportunity for collaboration between them. Worse yet, ISEs will recollect the data on a process step (time study, ergo, work instructions, etc.) each time that data is needed for a specific function as opposed to referencing and maintaining a common process database. If we are not saving, sharing, and maintaining process data is that because it is not inherently valuable? And if the data does not have value, what does that say about our profession that recollects that data time and time again? As the founders of efficiency, how has the process of performing industrial and systems engineering become so inefficient?

While this view may seem degrading at first, when looked at through another lens, such a view is exhilarating because of the potential that exists in improving the way we, as ISEs, practice our profession. ISEs have contributed greatly to the efficient operation of our factories and businesses, and now is the time to direct these skills inwardly and become the imaginary engineers that IEs are often called by other engineering professions. Imagine if ISEs had efficient and effective ways to implement their industrial and systems engineering skills sets, across multiple industrial and systems engineering disciplines, leveraging the same process step database across the organization. Imagine if the ISEs took the same approach to building and managing the Bill of Process (BOP) as the mechanical and electrical engineers take with the Bill of Materials (BOM). Such a change is possible, and we are on the cusp of that emerging frontier when taking an enterprise-wide, technology-enabled BOP approach.

8.2 Creating, Managing, Maintaining the BOP Effectively

The BOP is more than just a Process Routing. As stated by Littlefield in *The Evolution of MOM and PLM: Enterprise Bill of Process*, the BOP encompasses everything known about how to manufacture the product represented by the BOMs. It includes the plants (layout), resources (machines and tooling), work instructions, ergonomics assessments, quality plans, and process configuration rules.[1] Like a BOM, a BOP needs to be constructed, managed, and maintained. This effort takes collaboration which requires some common definitions.

The fundamental component of a process is the activity (often called a task) which represents the smallest amount of movable work. This definition is important because it represents the fact that Activities can be relocated to different people or machines. As shown in Figure 8.1, this relocation requires reaggregation at the new location, which is often performed automatically. Attaching attributes of time, tooling, ergonomics, quality, and part consumption to an activity ensures that this information is available for aggregation, reconciliation, and reporting in the future. For example, if you drag and drop five Activities to an assembly line workstation, you now can aggregate (compile) the total time for the operator, the total ergonomic risk, the quality control plans, and even the shop floor work instructions. Often these engineering functions are those which take an engineering department the most time, require the least skill, and generate the greatest error.

Of course, this means that all members of the Process Engineering team are collaborating around these common definitions of work and that new definitions of work (Activities) are only created when a unique task needs to

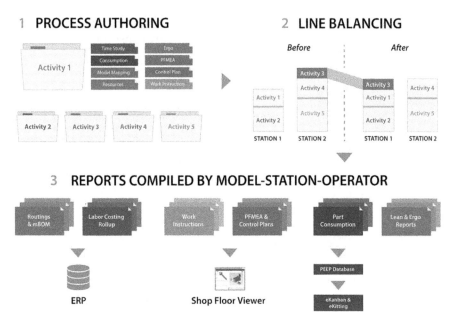

FIGURE 8.1
Activity assignment and compilation in the BOP.

be performed, otherwise existing Activities should be reused. This is identical to the benefit of reusing existing parts in a new BOM instead of creating new fasteners and brackets for every new product. Reusing Activities across the facility greatly reduces the redundant process engineering effort and enhances the quality of the data in the system and the documentation and decisions which come from it. In fact, without Activity reuse, a BOP is cost-prohibitive to create and maintain.

In order to reduce Activity proliferation, it is critical to Model and Option code the activities so that they can be referenced into the Process Routings created from the BOP. Figure 8.2 shows an example of Model Option code applied to Activities. This simple concept borrowed from BOM management means that we can have a master BOP for a department or assembly line which contains all relevant Activities which are currently performed. Since each Activity knows what Models and Options it is appropriate for, the Process Engineer can automatically filter the master BOP into an order-specific Process Routing, and generate reports and documentation for time, tooling, instructions, and quality on demand. Another feature is the ability to use "Where Used" queries on Activities to understand what all Routings (Model and Option families) this process task is associated with. Therefore, just like the need to understand what products are affected by this BOM component change, we can understand what all products are affected by this process or tooling change.

Most importantly, a common definition around Activities allows the ergonomist to reference the times and task descriptions from the time study people or the standard operating procedure (SOP)/work instruction authoring person to reference the time, tooling, and part consumption attributes

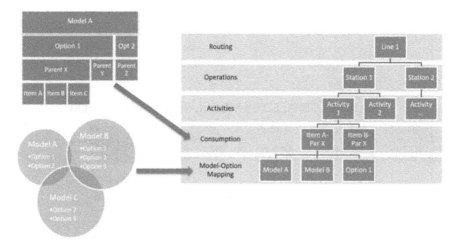

FIGURE 8.2
Model and option mapping to make activities reusable in many routings.

entered by other engineers. Collectively, this process information will be more accurate, more complete, and of much greater value to the organization than the set of disconnected, incomplete, and out-of-date spreadsheets which Industrial Engineers commonly adhere to today. With a complete and accurate BOP, organizations can make quicker and more accurate decisions on launching new products and processes, adjusting production volumes, or justifying and implementing process improvements. In short, companies will bring new products to market sooner, with less cost, and better quality.

8.3 Visualizing the Product and the Process in a Collaborative Way

Video and 3D product models are two relatively new technologies which enhance the quality of the process steps being defined and documented. Most importantly, both technologies greatly enhance the BOP usage and value throughout the organization and, thus, the perceived value of the ISE community responsible for creating and maintaining them.

8.3.1 Videos of the Process

To truly maximize the value ISEs can bring to their organizations through the Activities mapped to the BOP, every organization must have a video associated with every process Activity maintained. That is a strong statement and is often in conflict with rules in the organization prohibiting video, but those rules MUST change. Yes, this technology is just that important. If a picture is worth 1,000 words then a video is worth millions.

Videos can be used to create or verify observed time studies or benchmark predetermined ones, they show the workplace layout, the tooling used, the operator movements necessary for work instructions or ergonomic studies, the material handling equipment, the lighting, the noise and interactions with other operators and machines, and so much more. Collectively, the video IS the visualization of the BOP this chapter seeks to define.

Creating videos is easy but using them and managing them effectively takes help from technology. Most importantly, videos can capture many activities, and they can do so over multiple operator cycle times which may comprise of many Activities each. Figure 8.3 shows how one video can be comprised of many activities. Thus, the video recording needs to be easily segmented and cataloged by the engineer. This can be done by breaking the video into individual files or more appropriately associating file and time stamp links from a video file to Activities. In either case, each Activity should map to at least one video recording segment and perhaps several segments which might represent multiple observations (observed time study). The default

FIGURE 8.3
Observed multi-cycle activity time studies from a video.

video linked to each Activity will be the one referenced for work instructions, ergonomic studies, and predetermined time studies. Any employee in the company should be able to quickly hyperlink to this video from the system where they view their activity list. Common locations include process routings, shop floor manufacturing execution systems (MES), and enterprise resource planning (ERP) systems. A hyperlinked video is a considerably more effective way to train and inform workers than traditional printed SOP documents with photos.

Finally, once Activities have been reassigned to operators in the factory, the line balancing, software can recompile these Activity videos in sequence to create a video of the new operator operation.

8.3.2 3D Models Linked to the Process

Once you have established a link between the BOP and the BOM by having your process Activities consume your BOM components, you can leverage your 3D product models in amazing ways. A key enabler of this emerging trend is that 3D product model viewers are becoming very common in manufacturing organizations today. This is because most of the components manufactured and assembled today have a 3D model associated to them and the software and hardware cost of 3D viewing technology has come down substantially.

The most basic use of this 3D integration is the ability to view components in the BOM when you are consuming them by the process. Often this means simply selecting a hyperlink on a component to view that component; however, more advanced systems allow you to select components from a 3D assembly and drag and drop them to the process Activity that is consuming them. Then you can view a consumption-color-coded 3D assembly and see which components are consumed and which are not.

Having a 3D component or assembly linked to a process Activity then makes creating and annotating work instruction documents considerably easier and accurate and is particularly helpful for new products where photos may not be available. In this situation, it can also be helpful if your system allows you to associate a 3D model link to each Model/Option-dependent component that your activity is consuming. Put another way, if your Activity assembles three different alternators, then you may wish to associate the three different alternator models to the Activity so that the proper model image shows up when the Activity is referenced in a Routing or work instruction for a specific Model/Option combination.

Finally, for assembly-driven organizations, 3D Models can be very helpful when creating and visualizing assembly precedence diagrams and Yamazumi charts. Figure 8.4 shows a 3D process model that aligns with the associated Activities in the Yamuzumi chart, and Figure 8.5 shows a 3D model that correlates to the associated precedence chart. When editing product assembly precedence in the chart view, you can see the product automatically assemble itself. Simply selecting the Activity from the precedence map causes only the components assembled up to that Activity to display and will show the components in the selected Activity in a different color. With this approach, the assembly precedence can be visually validated.

Linked Yamazumi charts work in much the same way. Engineers can select a tile in the chart which is associated with an Activity and see the

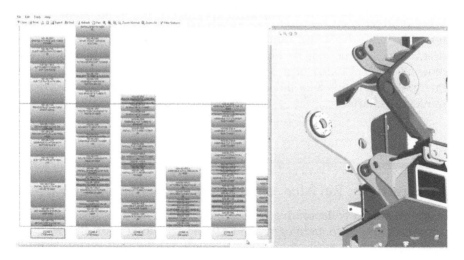

FIGURE 8.4
3D product model linked to a Yamazumi line balance chart.

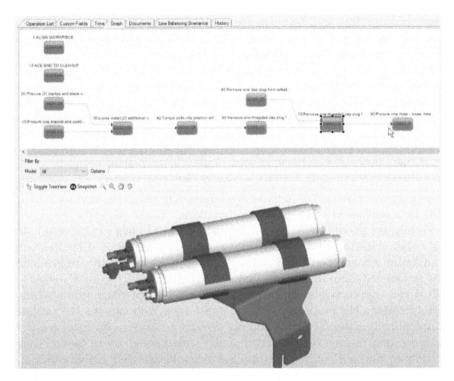

FIGURE 8.5
Assembly precedence diagram linked with 3D product model.

visual image of the assembly as of that specific Activity. Moving Activities around between stations (drag and drop the tiles) will then change the view of the product at each station. Engineers can then snapshot these product assembly views from the Yamazumi editor to create images for Activity-based work instructions or record the Yamazumi query (tile by tile) and generate a short video of a 3D virtual assembly for that operator. This emerging trend adds a lot more organizational value to creating and using Yamazumi charts.

8.4 Effectivity, Revision, and History Control in the BOP

In any managed collaborative environment, it is critical to have rules and controls to provide data integrity and user accountability. Just like with the BOM management technologies which have extended from simple spreadsheets through to product data management (PDM) and product life cycle management (PLM) environments, BOP databases need similar tools.

Now that we are managing, maintaining, and collaborating around Activities in the same way we do with components of the BOM, revision and effectivity can be applied.

8.4.1 Revision Control

Revision control is often a labeling technique to quickly understand that a change to this shared Activity has occurred. Often this can be coordinated with multiuser check-in and check-out mechanisms to ensure that multiple users are able to edit or view within the same BOP without corrupting the data between themselves. Additionally, workflow systems can be incorporated to ensure that revisions are reviewed and/or approved prior to release.

8.4.2 Effectivity Dates

Effectivity date range coding of the Activities works very similar to how routings and components can be effectivity date managed within ERP systems. The engineer simply defines a date (or event, or product series) where the company should transition from using the old Activity to the new one. Effectivity dates enable a common process plan BOP to be used in an environment where Activities referenced by the BOP are changing daily.

8.4.3 History Tracking

Finally, history tracking is the police of collaboration. It ensures that actions made by users of the BOP system are tracked and searchable so that should problems arise, accountability can be made, training enhanced, and mistakes reduced or at least quickly corrected.

8.5 Mapping the BOP and BOM and Reconciling the Process to the Product

Perhaps the most valuable benefit of a BOP which contains reused Activities is the ability for those Activities to consume (map to) the components (or machined features) in the BOM for which they are responsible. For example, a task to assemble an alternator onto a car may involve the alternator, a bracket, and perhaps a few bolts and washers. If the Process Engineer defines the Activity as consuming those parts from the BOM, then it is now possible to ensure (reconcile) that all Activities in the BOP consume all components in the BOM—no more and no less. This is critically important when large and complicated assemblies are launched to the production floor and the engineer finds out that the work instructions do not match the product they are producing and the parts in each station do not match what the operator needs.

Performing this consumption mapping process requires a lot of manual effort or some very intelligent software and a well-defined collaboration process.

1. First, your Activities must use the same Model and Option configuration rules as your BOM and your BOP, and the BOM must be defined for the same area of production (i.e., Assembly line 6). This way you can be certain that the components assembled in that area are assembled with the processes in that area.

2. Second, your Activities must be able to consume part numbers by Model and Option code. For example, if you have an Activity that assembles an alternator on a car, then this Activity could apply to four different alternators, depending on which of the four option codes mapped to that Activity are being referenced. Therefore, the engineer needs to specify all four alternators as consumed by the Activity and map the correct Option code to each of those mutually exclusive consumption events. Luckily, there is software available which can read a configured BOM and automatically populate the part consumption to configured Activities in the BOP. (Pretty cool huh?)

3. Third, we need to reconcile the consumption of components by an Activity to the BOMs for which that Activity consumes components. All BOMs in the master BOM mapped to all Routings in the BOP.

4. Finally, we can address errors arising from a failed BOM–BOP reconciliation until we achieve a resolution good enough to be launched on the production floor.

8.6 Industrial and Systems Engineering Technology Enablement—PLM–ERP–MES

For ISEs to bring systemic change to the way they perform industrial and systems engineering as described in the sections above, the means to do so must exist. Just as the automobile becoming a more efficient and effective means of transportation could not exist before the invention of the wheel and the internal combustion engine, the means for ISEs to manage data at the smallest process step and then share that data across multiple engineering tasks cannot be done without the use of PLM and MES. Today, many organization try to utilize ERP as the means to process all their data and thus have all data go into and flow from the ERP system. However, ERP systems are great at showing transactions that have happened (i.e., an accounting function) and for making future predictions about what parts and materials are going to be required (i.e., a purchasing function) but are not the right tool for the design or execution functions in an organization.

ERP systems are designed to integrate the information between manufacturing, inventory management, sales, accounting, service, and the customer relationship management (CRM) software. The key objective of ERP is to provide a high level of integration necessary to efficiently synchronize these functions so that the organization can make effective high-level decisions more responsively. In particular, the synchronization of sales to manufacturing, inventory, and accounting can have a profound impact on ensuring that the organization is delivering the "right" products to the "right" customers at the "right" time.

In this capacity, ERP systems are planning applications that are responsible for ensuring that parts, people, and tooling are available to produce what customers want to purchase, and the ERP system's primary "customer" is upper management who needs access to this information in a quick and coordinated manner.

PLM systems are essentially engineering design and which provide the manufacturing "Plans" for the ERP to execute (i.e., the design function), whereby MES applications are shop floor execution applications which functionally implement the manufacturing "Plans" dictated by the ERP system's "Build Schedule" (i.e., the execute function).

ERP systems are not the right environment for execution, and ISEs need the tools to operate in an execution-based environment. Figure 8.6 shows the data flow between these various tools that enable ISEs to complete their roles and responsibilities on a platform that enables efficiency and effectiveness.

To better understand this graphic and the way in which ISEs can apply industrial and systems engineering principles to the practice of industrial and systems engineering itself, understanding the foundation for the way in which these platforms interact is important.

As seen in the PLM, MES, and ERP data flow diagram (Figure 8.6), the majority of data required between PLM and MES applications are detailed "execution-oriented" and of no value to the ERP system. As such, attempting to integrate all of this information within ERP creates a great deal of complexity and cost without any benefit. In fact, the significant amount of data that would need to be imported into and out of ERP and associated to ERP objects would result in an unnecessary overall system performance reduction to all users.

In particular, the Routing and Tooling information required for PLM and MES is substantially more detailed than what is required by the ERP. Most ERP implementations define a routing in terms of Operations that require a composite time, whereby in PLM systems, Operations are defined from Activities which are defined by Worksteps that often are comprised by Elements.

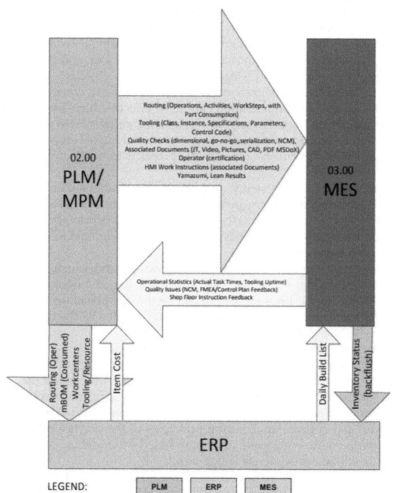

FIGURE 8.6
Data flows—PLM–MES–ERP.

Each of these process detail levels may contain from 2 to 100 records, and therefore, a simple 1 Operation record in ERP could represent from at least 10, to several hundred thousand records in PLM. This similar analogy applies to MES, whereby Activities and Worksteps form the backbone of a process description to an operator and elemental data is defined in machine code numerical control (NC), programmable logic controls (PLC), etc. to the tooling. Of course, PLM and MES systems handle this complexity via configuration rules so that this process detail needs to only be defined once but referenced in hundreds or thousands of different Routings. These configured process details at the Activity and Workstep levels also require sophisticated engineering change procedures, involving version, platform, or serialized effectivity, in a manner far more detailed than what ERP systems typically manage, and their users are experienced with.

As such, managing this data, detail, and complexity among a focused group of engineers in the PLM and MES applications, and their associated integration, will result in a more organizationally effective approach than would be gained by routing this workflow through ERP.

8.7 Within the PLM—Manufacturing Process Management

Within the PLM lives the manufacturing process management (MPM) system. The objective of MPM is to define the process, manage multiple versions of it (i.e., prior versions, active versions, and future planned alternates), and export the "Active" process with Effectivity attributes (i.e., serialized, version, or platform) to MES and PLM systems for Execution. Figure 8.7 shows the data flow within an MPM system.

The typical workflow for a new assembly or the more typical engineering change to an existing assembly looks as follows:

1. Engineering change order (ECO)/manufacturing change order (MCO): Receive ECO and create MCO and define change Effectivity.
2. Manufacturing BOMs (mBOM): Author or edit an associated mBOM.
3. mBOM Downstream Data: Electronically identify the associated downstream manufacturing data affected by the mBOM changes. Best accomplished via associative data linkages.
4. Process Engineering: Author or edit all associated routings and process objects (i.e., operations, activities, elements—includes time estimation).
5. Tooling: Evaluate tooling requirements (i.e., jigs, fixtures, automation code development [NC/coordinate-measuring machine (CMM)/ Robotic/Lightboard/PLC]).
6. Work Instructions: Define work instruction notes and associated documents and images.
7. Line Balancing: Evaluate work assignment to stations and operators (i.e., may involve line balancing/man–machine analysis). This subsequently results in the assignment of parts and tools to those stations.
8. Transactional Requirements: Define transactional requirements (inventory backflush, inventory/PFEP triggering, safety equip

specification, quality [FMEA/Control Plan/Statistical Process Control (SPC)/nonconformances (NCM), genealogy, tooling details]).

9. Publish mBOM and routing to ERP.

10. Publish instructions (with associated process, part, tooling data) to MES.

11. Publish part consumption changes to PFEP.

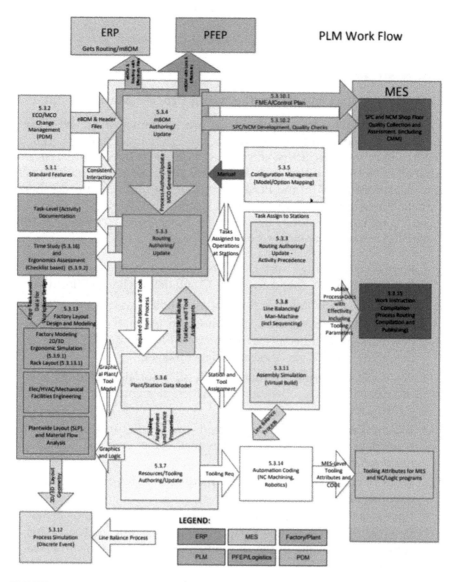

FIGURE 8.7
MPM workflow.

By utilizing an MPM system and the process outlined previously, changes to the design which are triggered by an ECO update the engineering BOMs which can then be compared against the existing mBOMs and an associated MCO can be issued to change the materials needed in the manufacturing process to accommodate the engineering change. Within the MPM system, when the mBOM changes, the associated activities being performed in the manufacturing/assembly operation are flagged to be updated accordingly, which then triggers updating the required tools, work instructions, line balance, etc. Rather than in a traditional model where each of these operations has their own data set which must be updated independently (often by different individuals), in the MPM framework the change is cascaded through the MPM system. This mitigates the room for error in which one part of the engineering subset is updated with the correct information, let's say the work instructions, but another area is missed, let's say the tooling requirements are not updated accordingly, leading to (in this example) a set of work instructions that the operator does not have the correct tools to perform. When errors such as this are mitigated, it allows for significant improvements in being able to implement engineering changes into production. With best-in-class MPM and MES systems, before a change is ever implemented, the data model for the change can be reconciled (similar to how accounting can reconcile the books) to ensure the product, process, plant, and resources are all updated and accurate for the change to effectively take place.

8.8 Case Studies

While many organizations are still using separate excel spreadsheets or making expensive modifications to their ERP systems to force that system to perform execution functions that the ERP was not designed to do, others are on the emerging frontier and have started to implement best-in-class PLM and MES systems, some out of necessity and others by choice.

One such example was at a large appliance manufacturer, who lost one-third of its workforce in 1 day due to legislative changes. The plant was set to run at a given line rate, and without one-third of the workforce, the plant would be forced to shut down unless the line rate could be adjusted accordingly. Adjusting the line rate requires rebalancing the line and updating work instructions, ensuring parts are sent to the correct stations with the rebalance, moving tooling accordingly, etc. Under traditional operating practices, such changes would not have been feasible in a short period of time given the vast magnitude of tasks and activities that engineers would need to update to adjust the line rate accordingly. However, this organization had previously moved their data to a best-in-class MPM and MES system, where the activities were associated to operations and routings, and thus, relatively few data changes were required in order to adjust the line rate and subsequently update all associated tooling requirements, work instructions, ergonomic assessments, etc. Given this, the plant was able to adjust the line rate and continue to production with the decreased workforce. In the words of the plant manager, being able to quickly adjust the line rate with the best-in-class MPM and MES system saved the plant.

In another example, a large agricultural/industrial equipment manufacturer typically took 2 months from the release of a new product to the shop floor to where the shop floor was producing at the production target. Often the reason it took this long to reach the production target were errors found after the initial launch where information had not been updated correctly so the new product launched with missing information (e.g., an inaccurate work instruction, tooling in the wrong location, process times that were different in the line balance model than in the time study). The organization moved to using the best-in-class PLM system but had plans to continue to use traditional printed work instructions and not deploy the MES system initially, yet ordered tablets to be able to utilize electronic shop floor work instructions through the MES system in the future. On the day of the product launch to the manufacturing floor, the production team discovered that the printed work instructions printer correlation settings had been set incorrectly, and the printed work instructions were useless as a result. The lead time to reprint all the work instructions was greater than a week. So, the team flipped on the MES and deployed the tablets to the floor in 1 day, thus delivering the correct work instructions to the correct stations. As a result of the work instructions being electronic, with a mechanism for the operators to provide feedback on incorrect work instruction information through the tablets, shop floor workers were able to flag any errors or changes in the work instructions back to the engineering team in the first days of production, and by the end of the week, all of the work instruction content was correct. Having the work instruction information be correct within the first week as a result of the operator feedback led to the company hitting their production target the fifth day after product launch, which had typically been a 2-month process, and the production manager was thrilled as such a quick deployment resulted in $2,000,000 of savings to the organization.

8.9 Imagining and Creating the Emerging Frontier

These examples show the power of taking the best practices of industrial and systems engineering and realizing not only the individual value of each subset of industrial and systems engineering, but the synergy when information can be shared across industrial and systems engineering disciplines through an enterprise-wide, technology-enabled BOP approach. When an Activity can be updated, and those updates apply across all associated BOPs and reconcile to the BOMs, the effort required to perform industrial and systems engineering functions is drastically reduced while at the same time increasing the accuracy of the results. Such an approach lives up to the true intent of the efficiency and effectiveness of industrial and systems engineering. As more academic institutions teach the value of how the different subsets of industrial and systems engineering work together, and as more organizations adopt best-in-class PLM and MES systems that allow them to fully realize the benefits of a cross functional industrial and systems engineering approach, the emerging frontier presented here will have transformational impact on society and the industrial and systems engineering profession. ISEs will be imaginary engineers not in the derogatory sense but in the sense that we imagined the future, and we created it.

Reference

1. Littlefield, M, The Evolution of MOM and PLM: Enterprise Bill of Process, 2012, http://blog.lnsresearch.com/bid/141670/the-evolution-of-mom-and-plm-enterprise-bill-of-process.

Section II

Engineering Applications and Case Studies

9

Remarks on Part II—Engineering Applications and Case Studies

Harriet B. Nembhard
Oregon State University

Elizabeth A. Cudney
Missouri University of Science and Technology

Katherine M. Coperich
FedEx Ground

In Section I, distinguished practitioners and academics discussed models for successful collaboration between industry and universities as well as methods for developing organizational initiatives and methods for increasing motivation and ethical behavior. These collaborations highlighted the need to explore the infrastructure, methods, and models for integrating academia and practice. Section II of this book provides case studies and applications to illustrate the power of the partnerships between academia and practice in industrial and systems engineering.

Chapter 10 presents a model for a maintenance strategy as a value-generating action, rather than from a cost-centric approach for civil infrastructures. The proposed approach is value driven, unlike existing maintenance planning models that focus on minimizing cost. This study illustrates the need for academics and practitioners to change paradigms from a cost focus to a value focus.

The benefits of collaboration between industry and academia are also highlighted in Chapter 11—this time with respect to emergency management and how it has become an industry. In particular, the facets of industrial and systems engineering and how they are beneficial to growing the process maturity in emergency management are discussed with key aspects centering on workforce development and training and maturity models.

Very small entities are organizations with up to 25 persons and are a growing section of the global economy, particularly with respect to systems and software developers. The need to develop a standard and qualify these organizations is discussed in Chapter 12. Barriers to entry in international business sectors can be removed through the development of these standards for women and minorities. Further, considerable opportunities for Industrial and Systems Engineering projects through partnerships with very small entities can also remove these barriers.

Chapter 13 provides a multidisciplinary and collaborative effort to applying Industrial and Systems Engineering approaches to combat human trafficking, a growing criminal economy. The counter-trafficking research uses visualization tools and real-time data sources to identify victims and discover trafficking networks. Mathematical models are also proposed to allocate resources to disrupt the human trafficking supply chain.

Technology commercialization in manufacturing process design is discussed in Chapter 14. Manufacturing process design involves the engineering activities necessary by manufacturing engineers to support product development. A framework is presented for manufacturing design that involves determining process requirement, selecting appropriate processes and machine tools, and evaluating the detain based on process-based cost modeling.

A virtual to reality big data methodology is proposed in Chapter 15 along with a case study at a manufacturing facility. As big data continues to become a concern for more and more companies with massive advancements in technology, it is important to understand the usefulness of gathered data and how it can be used to improve operations. A virtual to reality big data methodology is proposed using artificial intelligence/machine learning methods to handle the data.

Real-time monitory and control of cyber-physical systems is the focus of Chapter 16, specifically with respect to biomass-based energy production. The use of cyber-physical systems is discussed to increase the intelligence across the biomass-to-bioenergy supply chain to overcome the challenges associated with multiple spatial and temporal scales.

Finally, Chapter 17 discusses how productive and continuous initiatives can be employed using a change management model to introduce adaptable, deliberate, and systematic cultural change in an organization. The proposed Meta Change Model was deployed at a multinational brewery company to introduce Lean Six Sigma starting in financial and support areas, and then institutionalized internationally.

10

Value-Based Maintenance for Deteriorating Civil Infrastructures

Seyed A. Niknam
Western New England University

Alireza Jamalipour
The Connecticut Department of Transportation

CONTENTS

10.1 Introduction

Humans are more dependent upon civil infrastructures than ever before. All civil infrastructures deteriorate over time. America's aging and under-performing infrastructure not only tarnish the nation's prosperity but also becomes more dangerous and costly for the citizens. The American Society of Civil Engineers reports a comprehensive assessment of infrastructures every 4 years (ASCE 2017). According to the recent American Society of Civil Engineers' 2017 Infrastructure Report Card, the nation's infrastructure is in fair to poor condition causing the loss of $9 a day for each American family. This report distinguished 16 different categories of civil infrastructure including transportation, drinking water systems, energy, and waste management. Throughout this chapter, we focus on bridges that represent one key element of any transportation system. Table 10.1 shows the grades for the transportation section: bridges and roads (ASCE 2017).

TABLE 10.1

Transportation Section Grades in the Infrastructure Report (2017)

	1998	2001	2005	2009	2013	2017
Bridges	C–	C	C	C	C+	C+
Roads	D–	D+	D	D–	D	D

The average age of the United States' 614,387 bridges increased to 43 years in 2017. Considering the age of bridges, as shown in Figure 10.1 (ASCE 2017), one can realize that the transportation system is in urgent need for beneficial opportunistic maintenance strategies. It must be cautioned that the bridges constructed in the period of the 1950s–1970s have the anticipated service life of 50 years. Moreover, the age of the 503 bridges failed between 1989 and 2000 ranged from 1 to 157 years, with an average of 52.5 years (Wardhana and Hadipriono 2003).

Almost 9% of the nation's bridges are structurally deficient. Figure 10.2 shows the congressional map of deficient bridges in the United States based on the data released by the Federal Highway Administration (FHWA) (ARTBA 2018).

In the state of Connecticut, there are 338 structurally deficient bridges, i.e., 8% of the total 4,214 bridges. The top ten most traveled structurally deficient bridges in the state have a daily crossing of over 1.2 million vehicles, with an average age of 58.6 years. As it currently stands, 174–188 million trips are made daily across structurally deficient bridges in the United States. To our surprise, the average age of structurally deficient bridges is almost 65 years. Furthermore, in 2016, more than one in eight of the nation's bridges were

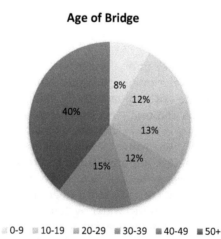

Age of Bridge

0-9   10-19   20-29   30-39   40-49   50+

FIGURE 10.1

Age of bridges in the United States (ASCE 2017).

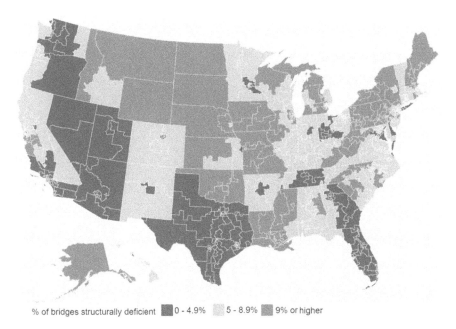

% of bridges structurally deficient ■ 0 - 4.9% ░ 5 - 8.9% ▓ 9% or higher

FIGURE 10.2
Congressional map of structurally deficient bridges (ARTBA 2018).

functionally obsolete which means they fail to satisfy current standards or traffic demand.

Despite the advances in life-cycle management, few attempts have been made to apply life-cycle models in design and assessment codes of infrastructures. Performance assessments of bridges rely merely on visual inspection. In visual bridge inspection, major components (i.e., the deck, superstructure, and substructure) are rated. Poor condition in one of the major components will result in categorizing the bridge as structurally deficient (Jamalipour, Niknam, and Cheraghi 2017). Figure 10.3 shows typical compiled bridge inspection data.

Poor conditions, normally caused by deterioration, entail the need for preservation actions and limit the bridge usage, particularly for heavy truck traffic. According to the FHWA, the bridge preservation is defined "as actions or strategies that prevent, delay or reduce deterioration of bridges or bridge elements, restore the function of existing bridges, keep bridges in good condition and extend their life" (FHWA 2011). Bridge preservation starts with an inspection in order to assess the physical condition of the bridge components. The inspection will determine the degree of impairment and load-carrying capacity in the existing condition. The inspection is more deep seated for fracture critical members and foundation members under water. It is evident that the bridge inspection procedure is dependent on the structure type and the material of the bridge's longest span. According to the American

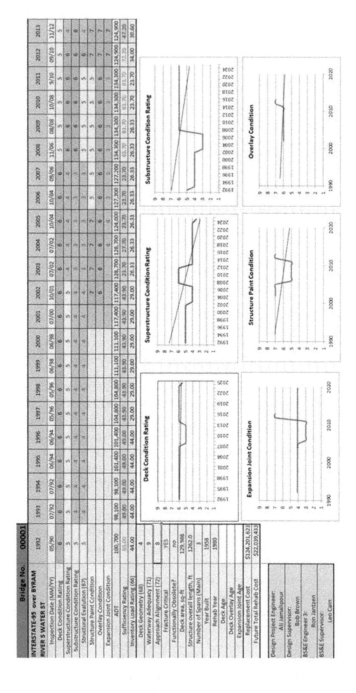

FIGURE 10.3
Typical compiled bridge inspection data.

Association of State Highway and Transportation Officials (AASHTO), bridge inspection should be carried out at least every 2 years, and the data collected on structure inventory and appraisal should be reported to the National Bridge Inventory (NBI) maintained by the FHWA. The inspection should be followed by evaluation and maintenance planning. Part of the evaluation is using the database of bridge network to predict the rates of deterioration and forecast major bridge rehabilitation cost (Jamalipour and Niknam 2017).

The main obligation of transportation authorities is to preserve the safety of structures and to improve public services at minimal cost. In bridge preservation process, the actual values of bridges are not taken into account. Focusing on value perspective, this study intends to investigate the application of value-based maintenance strategies for deteriorating civil infrastructures. Our emphasis will be on maintenance strategies for highway bridges.

10.2 Life-Cycle Maintenance of Deteriorating Infrastructures

Life-cycle assessment of infrastructures is a popular issue to researchers (van Noortwijk and Frangopol 2004; Frangopol, Dong, and Sabatino 2017). Affected by deteriorating events, the performance of infrastructures varies over time as shown in Figure 10.4 (Morcous 2006).

To appraise the entire life of structures, performance prediction models have become an integral part of life-cycle management. Over the life of a system, there are many local failures with different modes and occurrence rates. Performance indicators for structural systems should provide the necessary information to predict the occurrence of local failures and evaluate their damage propagation to prevent the system from a global failure. In this respect, there are a wide range of aleatory and epistemic uncertainties

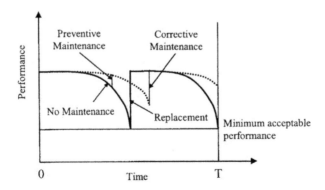

FIGURE 10.4
Performance versus time (Morcous 2006).

related to infrastructure performance ranging from material properties and load distribution to the occurrence and magnitude of natural hazards.

There are a variety of performance indicators used in the literature to capture uncertainties:

- Reliability-based indicators use time-variant failure probabilities and, therefore, deal with load and resistance uncertainties. Reliability-based indicators can be defined as an individual component or the whole system.
- Risk-informed decision-making takes into account failure probabilities and their consequences (Lounis and McAllister 2016; Ang and De Leon 2005).
- Sustainability indicators integrate environmental and social aspects of decision-making (Sabatino, Frangopol, and Dong 2015).

The most widely used measure in the evaluation of infrastructure performance is life-cycle cost. For this, numerous maintenance models have been developed on the basis of life-cycle cost minimization (Frangopol and Liu 2007; Wu, Niknam, and Kobza 2015; Frangopol, Dong, and Sabatino 2017). The expected life-cycle cost of an infrastructure consists of several elements such as initial costs, maintenance costs, and expected failure costs. The choice of construction materials is, indeed, one major contributor to the expected life-cycle cost of bridges. Conventional carbon steel bridges normally have lower initial cost. On the contrary, corrosion-resistant steels, such as American Society for Testing and Materials (ASTM) A1010 steel, come with a higher initial cost. However, the life-cycle cost of conventional steels, i.e., repainting, is considerably higher than corrosion-resistant steels in low chloride environments (Okasha et al. 2012).

Maintenance costs can be divided into two major classes: direct and indirect costs. The former includes the cost of preventive maintenance (PM), cost of monitoring and inspection, and cost of repair. For a thorough cost analysis, there are three main classes of maintenance to be considered. Corrective maintenance, which is also called substantial, reactive, or essential maintenance (EM), deals with damage repair when the system can no longer function or provide a service. In this type of maintenance, performance level drops below a predefined threshold which creates the need for rehabilitation or repair. Rehabilitation refers to the activities that restore a bridge component to the required standards.

The second class is PM. According to AASHTO's Subcommittee on Maintenance, "preventive maintenance is a planned strategy of cost-effective treatments to an existing roadway system and its appurtenances that preserves the system, retards future deterioration, and maintains or improves the functional condition of the" (Hurt and Schrock 2016). Therefore, PM involves activities to maintain a consistent condition and to prevent or mitigate deterioration. The PM activities can be categorized into scheduled activities on a regular basis or response activities for issues identified during the bridge inspection. For example, due to deicing salts or marine environmental effects, steel bridges need scheduled maintenance such as cleaning bridge drains and scuppers. Those bridges may need response PM such as zone painting, spot repair, and overcoat for corroded elements or complete painting of the bridge after many years of service. Figure 10.5 shows the comparison of maintenance cost for PM and EM (Frangopol, Dong, and Sabatino 2017).

The third type of maintenance and the least popular type for bridge structures refers to a set of technologies such as strain gauges and acoustic emission, which provides sensory data as the base for condition-based health management. Extending service life and timely damage detection are the major goals in health monitoring of infrastructures (Nair and Cai 2010). In fact, to handle epistemic uncertainties, solutions must be found to reduce or eliminate incomplete information by means of effective health monitoring (Biondini and Frangopol 2016). Monitoring of infrastructures can be divided into two major classes: (i) global monitoring, i.e., for the entire system and (ii) local monitoring for specific damaged areas. Moreover, the duration of monitoring is classified into long term and short term mainly to update the integrity status.

In addition to the abovementioned direct costs, maintenance actions are normally associated with indirect environmental, social, and economic impacts. For example, bridge maintenance may cause traffic delays for transportation networks. The traffic delay would result in significant increase of bridge life-cycle cost depending on the bridge type, bridge location, average daily traffic (ADT), number of lanes, and capacity of the transportation network, among other factors.

In practice, ADT is time dependent and has an important influence on total indirect costs (Soliman and Frangopol 2015). Based on the sensitivity analysis of ADT and traffic-related costs, Kendall et al. suggest that excessive delays may cause a dramatic increase in life-cycle costs, and therefore, accurate assessment of road capacity and ADT is essential (Kendall, Keoleian, and Helfand 2008). The cost of time loss (C_{TL}) for users and goods is formulated as (Stein et al. 1999).

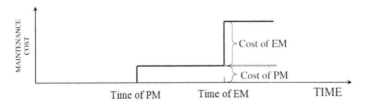

FIGURE 10.5
Maintenance cost.

$$C_{TL} = \left[c_w O_c (1 - T) + \left(c_c O_t + c_g \right) T \right] TL$$

where c_w represents the average wage per hour; c_c is the average hourly compensation for truck drivers; c_g denotes the time value of the goods transported in a cargo; T is the ratio of the average daily truck traffic to the ADT; O_c and O_t are the average occupancies for cars and trucks, respectively; and TL is the time loss, suggested by (Shiraki et al. 2007).

$$TL = (d)ADT \left(L(S_O - S_D)/S_O S_D \right)$$

where d is the duration of maintenance, L is the length of the traffic region, and S_O and S_D represent the unrestricted and restricted traffic speed, respectively. In addition, Kendall et al. proposed an approach to calculate the cost of the environmental impact and carbon dioxide emission as a function of ADT, L, and d (Kendall, Keoleian, and Helfand 2008). It is important to mention that indirect impacts of maintenance have been extensively discussed in sustainability-related publications (Sabatino, Frangopol, and Dong 2015; Bocchini et al. 2014; Penadés-Plà et al. 2016).

The expected failure cost has appeared in certain life-cycle models (Wu, Niknam, and Kobza 2015). Bridge failure is defined as the incapacity to perform as stated in the design and construction requirements (Wardhana and Hadipriono 2003). Hazardous and extreme events, floods in particular, are to be blamed for sudden bridge failures. The second major cause of bridge failure is overloading. For this, the inspection procedure should include load rating to determine the available live load capacity of structures. Deterioration, however, is not the major factor for a sudden bridge failure. Deteriorating mechanisms include but not limited to aging, wear due to loads and environmental conditions, concrete cracks and spalls, corrosion, fatigue, and hazards (e.g., blasts, earthquake, fire). Accumulating deterioration is the primary reason to turn a bridge to the condition of unserviceable. Excessive deflections in the superstructure and fatigue of tension members in steel truss bridges are the examples of accumulating deterioration.

Predicting the future maintenance requirements include the use of deterioration models. Such models would anticipate the influence of maintenance scenarios on the performance of the whole transportation network. As mentioned earlier, condition ratings are the prevalent outputs of performance assessment of bridges. For this reason, Markov chain models are the widely used stochastic models to predict the performance of infrastructures (Morcous 2006). Certain bridge management systems apply Markov chain models because of simplicity and computational efficiency at a network level. Nevertheless, the reliability of outcomes in Markov chain models is affected by limiting assumptions of state independence and discrete condition states.

10.3 Value-Based Maintenance

10.3.1 Value of Maintenance

For many years, the meaning of "value" has been the subject of discussion among economists. For this, there are many value-related theories and technical terms in economics and marketing (Doyle 2008). In recent years, customer-perceived value has gained much attention. In this context, value may be defined as a function of perceived benefits and perceived costs:

$$\text{Value} = \text{Benefits}/\text{Cost}$$

The equation is also presented in the other form as

$$\text{Value} = \text{Quality} + \text{Function}/\text{Investments}$$

Despite the fact that value is an intrinsic property of a system, it is very difficult to quantify a system's value. Generally, the quantification of the value is highly subjective and depends on the system usefulness defined by end users or customers. It is conceivable that "Maintenance has no intrinsic value" (Rosqvist, Laakso, and Reunanen 2009). Nevertheless, maintenance can be regarded as a value-adding initiative. Here, the important question is how we can quantify value-adding maintenance activities and apply those in maintenance planning. In practice, numerous factors contribute to the overall value of an engineering system; the most significant are the contribution to profit, market value, data significance, service to the community, satisfying regulations, and so on. In fact, maintenance actions will have both positive and negative impacts on the overall value of a system.

One significant impact of maintenance is the improved reliability of the system. The value of reliability improvement appears in the form of additional operational time and reduced probability of failure. Furthermore, it is important to keep in mind that the system reliability will not remain steady throughout the operational lifetime. However, questions still linger over how to identify and quantify reliability improvement during maintenance (Marais 2013). In a pioneering study, Saleh and Marais connected the financial value of reliability with system's net present value (Saleh and Marais 2006).

The value of maintenance can be perceived from the cost of failure. To understand the tremendous cost of failure for a major highway bridge let us look at the collapse of I-35W bridge (on August 1, 2007, in Minneapolis, Minnesota), which unfortunately resulted in substantial fatalities (Hurt and Schrock 2016). There is no way to quantify the profound cost of the lives lost in such events. Here, the economic costs and impacts need to be evaluated. The collapse of I-35W bridge limited the direct access to and from downtown Minneapolis for many commercial drivers and commuters. According to the analysis provided by the Minnesota Department of Transportation, unserviceability of the bridge cost the users $400,000 per day only for the increased travel time. The impact on local businesses due to reduced economic activity was around $113,000 per day. The total economic loss of the state was estimated around $60 million a year. Although the collapse happened due to problematic design, it is probably true to claim that the huge cost of failure was not considered in maintenance planning of I-35W.

The revenue-generating capability, i.e., provided service over time, is another way to quantify the value of maintenance (Niknam, Acosta-Amado, and Kobza 2017). This method is useful under the assumption that with no maintenance the system will reach the end of life. Based on this assumption, the net value of maintenance can be defined as the difference between the system revenue and the maintenance cost:

$$\text{Net value} = \text{Revenue} - \text{Costs}$$

The abovementioned net value over the life cycle can be used in the maintenance optimization function (Niknam, Acosta-Amado, and Kobza 2017). In essence, there has been little research in which the maintenance optimization function is based on maintenance value (Marais and Saleh 2009). Marais estimated the net present value for various system conditions and identified the optimal maintenance action through dynamic programming (Marais 2013). Obviously, maintenance actions come at a price. In practice, the planning horizon influences the type of value-optimal maintenance actions (Marais 2013).

To this end, it should be emphasized that while it is possible to estimate the monetary value of a system, it might be difficult to justify that value for maintenance planning. Rosqvist et al. outlined the major challenges in maintenance planning as clarifying objectives, identifying components to restore or repair, and deciding about maintenance tasks (Rosqvist, Laakso, and Reunanen 2009). From a value-based perspective, maintenance should be able to support the fundamental objectives (or values) of a system. Task planning deals with specifying detailed timing and needed recourses. Maintenance planning is dynamic, and therefore, the timing of maintenance has a great impact on the flow of service, i.e., customer-perceived values. In addition, value-driven maintenance will not be effective if there is no analysis of how much the condition of a system has improved by maintenance activities. Furthermore, any maintenance planning should be based on realistic system classification which takes into account significant factors, such as revenue-generating capability, public safety considerations (potential to cause loss of life or injuries), essential public service, ability to repair, and the possibility of partial failure, among others.

10.3.2 Value-Based Degradation Models

There has been little research on applying degradation data in performance prediction of bridges. In particular, the research area of value-based maintenance strategies for systems subject to degradation has been conspicuously lacking on published literature. In an attempt to develop a maintenance policy for systems with continuously degrading components, Bin et al. modeled components value as a function of a component reliability distribution (Liu et al. 2014). The concept of yield-cost importance was introduced to evaluate the shift in reliability distribution (i.e., the net value variation) and to determine the components that should receive the maintenance (Liu et al. 2014). In this research, the components in a series–parallel system are mutually independent, components are continuously monitored, and the degradation is not influenced by maintenance. These assumptions make it difficult to implement the model for bridge structures. In recent work related to value-based maintenance, an optimization model was developed to maximize the net value of imperfect degradation-based maintenance using the post-repair degradation level and the optimal interval of condition monitoring (Niknam, Acosta-Amado, and Kobza 2017). The model is based on monitoring the degradation data and performing imperfect maintenance when there is an indication of impending failures. Thus, two thresholds were defined: (i) the threshold between potential failure and normal state (X_P), and (ii) the threshold between functional failure and potential failure (X_F). Hence, X_P is the threshold that activates imperfect preventive repairs (PRs) and X_F is an indication of functional failure. Figure 10.6 shows a typical degradation with one imperfect repair (Niknam, Acosta-Amado, and Kobza 2017).

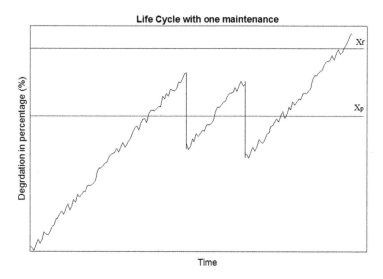

FIGURE 10.6
Typical degradation with imperfect repair.

In such system, which can be a bridge structure, a number of preventive cycles may happen until a failure is detected, i.e., the failure cycle. Therefore, maintenance activities have nothing to do with the first cycle and the failure cycle, i.e., no value-adding maintenance. Consequently, the life cycle is approximated by the expected length of preventive cycles. Hence, this model sounds promising to be implemented for bridge structures based on the assumptions that degradation is monotonic, the system is periodically monitored, and the system experiences only soft failures.

The model intends to maximize the net value of a system, which is the difference between the expected total cost of maintenance during a life cycle and the generated revenue. Hence, the life-cycle maintenance cost consists of the cost of failure (C_F), the cost of monitoring and inspection (C_M), a fixed cost of PR (C_S), and the cost of degradation reduction in a PR. To obtain the overall expected cost of PRs ($E[C_P]$), the expected degradation reduction after a repair ($E[R]$) is needed. Thus, assuming M is a known proportional constant, we have

$$E[C_P] = M(E[R]) + C_S$$

Then, the overall expected cost of PRs (TC) during all PM cycles can be obtained as

$$TC = C_F + E[N_P]E[C_P] + E[N_P + 1]C_M$$

Similarly, assuming that the amount of revenue generated in each preventive cycle (RV) is known, the value of the system can be defined as

$$V = RV\big(E[\text{Preventive cycles}]E[N_P + 1]\big)$$

Therefore, the generated revenue is dependent on the expected life-cycle length. This model suggests interesting results, which would not be readily obtained through cost-based models. It is clear that the risk of failure increases with longer monitoring intervals. For the same reason, the value-based strategy advocates more frequent monitoring. Furthermore, the model proposed that the optimal net value can be insensitive to the post-repair degradation level. It implies that one can reach the maximum maintenance value even without performing the highest level of repairs.

To further improve the value-based degradation models, one may consider the rate of degradation, accurate calculation of the degradation reduction, and revenue calculation due to partial failure. In the abovementioned models, an important shortcoming from the value perspective is neglecting the duration of PM. During the bridges PM, the public service is limited or interrupted, which is significant in value perspective. Moreover, the models assume that the only cause of system failure is degradation and, therefore, ignore the hazardous events that possibly impose longer maintenance period and more intensive repairs. In essence, the performance of the model in probabilistic life-cycle context highly depends on taking appropriate assumptions.

10.4 Discussion

It is worth to notice that unlike the economic value of precious assets, the value of an engineering system is not solid over time. Certainly, the value and price are not the same for engineering systems and the service they provide. This arises the need for moving beyond cost minimization models and considering the return of the investment in infrastructure. For any transportation system, bridges are a critical asset with tremendous values for owners and users. From the user's perspective, the value of a bridge relates to the satisfaction of their transportation needs.

Informed decision-making has become a common practice in design and life-cycle management on infrastructure. The various sources of uncertainties are one of the primary reasons to perform informed decision-making. Performance assessment and prediction are also at the center of informed decision-making. The success of both cost-centric and value-driven maintenance highly depends on the appropriate selection and calculation of performance indicators. In recent years, resiliency has been considered as another structural performance indicator. Resilience structures have the ability to adapt to changing conditions and recover from disruptions (Bocchini et al. 2014). The quantification of resilience can be done in deterministic or probabilistic framework, for individual bridges and bridge networks. The evaluation of the value of resilience structures is absent in the literature. In addition, there have been trends consisting of applying multi-objective optimization in order to assess conflicting criteria, which may provide more realistic bridge management plans.

As stated earlier, discrete condition states are the basis for Markov chain performance models. Taking a value perspective, it seems reasonable to use condition states to quantify system value in terms of flow service. Using degradation data in value-based models would bring more useful results. Technological advances in information systems have simplified the data collection and processing. Fruitless efforts are needed to bridge the gap between the theory and practice.

Maintenance strategy consists of defining the objectives, classifying the maintenance classes, and finally, selecting maintenance tasks. The strategy depends on the performance analysis of structures which can be done at the component level, for the whole structure or for a group of structures, i.e., a network. Keep in mind that bridges are the most vulnerable elements of a transportation network. In fact, risk and sustainability indicators have been applied for the performance assessment of bridge networks. Additionally, the performance analysis of bridge networks may use utility-informed decision-making to describe the relative desirability of maintenance policies for decision makers (Frangopol, Dong, and Sabatino 2017). Finally, to simplify the analysis at a network level, the focus may be shifted to major bridges vital to the transportation networks. A bridge structure that possesses one or more of the following characteristics is considered to be a major bridge: any bridge that crosses over a major river, any bridge deck with a square footage totaling over 100,000, bridges with movable part, any bridge with a unique structural design or complex geometry, and any bridge with detrimental effect to the transportation network.

10.5 Summary

All engineering systems fail or reach the end of their life. On this basis, particular attention should be paid to economic impacts of aging and deterioration of civil infrastructures. Maintenance is a primal necessity for deteriorating infrastructures. Maintenance tasks intend to secure functionality, detect and prevent failures, and restore a system to an operable state. Bridge maintenance requires abundant resources and appropriate planning. The main objective of any transportation agency is to determine, in advance, the time and budget needed for all necessary maintenance activities. However, unclear value-adding outputs led the decision makers to consider maintenance as a cost driver. Hence, value-adding aspects of maintenance is an important topic that justifies further discussion.

Following the infrastructure maintenance literature, it is conceivable that value-informed models may result in different, potentially, and more fruitful maintenance decisions. Despite various research efforts into applying value-driven maintenance, it is clear that there are still many issues that require further investigation. First and foremost, it is challenging to precisely quantify maintenance value since it is compounded by a combination of numerous factors. It would be easier to justify the investment on infrastructure if the benefits are explicitly quantified. It is relevant to mention that maintenance cost is always part of the value analysis. Therefore, value-driven maintenance planning provides more sophisticated system analysis.

Future focus in value-driven maintenance is about processing degradation data combined with possibility of hazardous events, identifying the value of partial failures, quantifying customer-perceived values, and value-based task planning.

References

Ang, AH-S, and D De Leon. 2005. Modeling and analysis of uncertainties for risk-informed decisions in infrastructures engineering. *Structure and Infrastructure Engineering* 1(1): 19–31. doi:10.1080/15732470412331289350. Taylor & Francis Ltd.

ARTBA. 2018. 2018 American Road and Transportation Builders Association Bridge Report. www.artbabridgereport.org/congressionaldistricts.html.

ASCE. 2017. American Society of Civil Engineers Report Card for America's Infrastructure. www.infrastructurereportcard.org/wp-content/uploads/2017/10/Full-2017-Report-Card-FINAL.pdf.

Biondini, Fabio, and Dan M. Frangopol. 2016. Life-cycle performance of deteriorating structural systems under uncertainty: Review. *Journal of Structural Engineering* 142(9). doi:10.1061/(ASCE)ST.1943-541X.0001544.

Bocchini, Paolo, Dan M. Frangopol, Thomas Ummenhofer, and Tim Zinke. 2014. Resilience and sustainability of civil infrastructure: Toward a unified approach. *Journal of Infrastructure Systems* 20(2): 4014004. doi:10.1061/(ASCE)IS.1943-555X.0000177.

Doyle, Peter. 2008. *Value-Based Marketing: Marketing Strategies for Corporate Growth and Shareholder Value.* Hoboken, NJ: John Wiley & Sons.

FHWA. 2011. Preservation Guide Bridge Maintaining a State of Good Repair Using Cost Effective Investment Strategies. www.fhwa.dot.gov/bridge/preservation/guide/guide.pdf.

Frangopol, Dan M., You Dong, and Samantha Sabatino. 2017. Bridge life-cycle performance and cost: Analysis, prediction, optimisation and decision-making. *Structure and Infrastructure Engineering* 13(10): 1239–57. doi:10.1080/15732479.2016.1267772. Taylor & Francis.

Frangopol, Dan M., and Min Liu. 2007. Maintenance and management of civil infrastructure based on condition, safety, optimization, and life-cycle cost. *Structure and Infrastructure Engineering* 3(1): 29–41. doi:10.1080/15732470500253164. Taylor & Francis.

Hurt, Mark A., and Steven D. Schrock. 2016. *Highway Bridge Maintenance Planning and Scheduling.* Oxford, UK: Butterworth–Heinemann.

Jamalipour, Alireza, and Seyed A. Niknam. 2017. Conceptual approach to proactively appraising and forecasting major bridge rehabilitation cost. In *ITS World Congress.* Montreal.

Jamalipour, Alireza, Seyed A. Niknam, and S. Hossein Cheraghi. 2017. Predicting highway bridge condition rating using markov models. In *IISE Annual Conference,* 362–67. Pitssburg, PA. https://search.proquest.com/openview/5e297a3a8aaa17746d09033b97be50bb/1?pq-origsite=gscholar&cbl=51908.

Kendall, Alissa, Gregory A. Keoleian, and Gloria E. Helfand. 2008. Integrated life-cycle assessment and life-cycle cost analysis model for concrete bridge deck applications. *Journal of Infrastructure Systems* 14(3): 214–22. doi:10.1061/(ASCE)1076-0342(2008)14:3(214).

Liu, Bin, Zhengguo Xu, Min Xie, and Way Kuo. 2014. A value-based preventive maintenance policy for multi-component system with continuously degrading components. *Reliability Engineering & System Safety* 132(December): 83–9. doi:10.1016/J.RESS.2014.06.012. Elsevier.

Lounis, Zoubir, and Therese P. McAllister. 2016. Risk-based decision making for sustainable and resilient infrastructure systems. *Journal of Structural Engineering* 142(9): 1–14. doi:10.1061/(ASCE)ST.1943-541X.0001545.

Marais, Karen B. 2013. Value maximizing maintenance policies under general repair. *Reliability Engineering & System Safety* 119(November): 76–87. doi:10.1016/J.RESS.2013.05.015. Elsevier.

Marais, Karen B., and Joseph H. Saleh. 2009. Beyond its cost, the value of maintenance: An analytical framework for capturing its net present value. *Reliability Engineering & System Safety* 94(2): 644–57. doi:10.1016/J.RESS.2008.07.004. Elsevier.

Morcous, George. 2006. Performance prediction of bridge deck systems using Markov chains. *Journal of Performance of Constructed Facilities* 20(2): 146–55. doi:10.1061/(ASCE)0887–3828(2006)20:2(146).

Nair, Archana, and Steve C. S. Cai. 2010. Acoustic emission monitoring of bridges: Review and case studies. *Engineering Structures* 32: 1704–14.

Niknam, Seyed A., Rolando Acosta-Amado, and John E. Kobza. 2017. A value-based maintenance strategy for systems under imperfect repair and continuous degradation. In *2017 Annual Reliability and Maintainability Symposium (RAMS)*, 1–6. IEEE. doi:10.1109/RAM.2017.7889689.

van Noortwijk, Jan M., and Dan M. Frangopol. 2004. Two probabilistic life-cycle maintenance models for deteriorating civil infrastructures. *Probabilistic Engineering Mechanics* 19(4): 345–59. doi:10.1016/J.PROBENGMECH.2004.03.002. Elsevier.

Okasha, Nader M., Dan M. Frangopol, Fred B. Fletcher, and Alex D. Wilson. 2012. Life-cycle cost analyses of a new steel for bridges. *Journal of Bridge Engineering* 17(1): 168–72. doi:10.1061/(ASCE)BE.1943–5592.0000219.

Penadés-Plà, Vicent, Tatiana García-Segura, José Martí, and Víctor Yepes. 2016. A review of multi-criteria decision-making methods applied to the sustainable bridge design. *Sustainability* 8(12): 1295. doi:10.3390/su8121295. Multidisciplinary Digital Publishing Institute.

Rosqvist, Tony, Kari J. Laakso, and Markku O. E. Reunanen. 2009. Value-driven maintenance planning for a production plant. *Reliability Engineering & System Safety* 94(1): 97–110. doi:10.1016/J.RESS.2007.03.018. Elsevier.

Sabatino, Samantha, Dan M. Frangopol, and You Dong. 2015. Sustainability-informed maintenance optimization of highway bridges considering multi-attribute utility and risk attitude. *Engineering Structures* 102(November): 310–21. doi:10.1016/J.ENGSTRUCT.2015.07.030. Elsevier.

Saleh, Joseph Homer, and Karen Marais. 2006. Reliability: How much is it worth? Beyond its estimation or prediction, the (Net) present value of reliability. *Reliability Engineering & System Safety* 91(6): 665–73. doi:10.1016/J.RESS.2005.05.007. Elsevier.

Shiraki, Nobuhiko, Masanobu Shinozuka, James E. Moore, Stephanie E. Chang, Hiroyuki Kameda, and Satoshi Tanaka. 2007. System risk curves: Probabilistic performance scenarios for highway networks subject to earthquake damage. *Journal of Infrastructure Systems* 13(1): 43–54. doi:10.1061/(ASCE)1076–0342(2007)13:1(43).

Soliman, Mohamed, and Dan M. Frangopol. 2015. Life-cycle cost evaluation of conventional and corrosion-resistant steel for bridges. *Journal of Bridge Engineering* 20(1): 6014005. doi:10.1061/(ASCE)BE.1943–5592.0000647.

Stein, Stuart M., G. Kenneth Young, Roy E. Trent, and David R. Pearson. 1999. Prioritizing scour vulnerable bridges using risk. *Journal of Infrastructure Systems* 5(3): 95–101. doi:10.1061/(ASCE)1076–0342(1999)5:3(95).

Wardhana, Kumalasari, and Fabian C. Hadipriono. 2003. Analysis of recent bridge failures in the United States. *Journal of Performance of Constructed Facilities* 17(3): 144–50. doi:10.1061/(ASCE)0887–3828(2003)17:3(144).

Wu, Fan, Seyed A. Niknam, and John E. Kobza. 2015. A cost effective degradation-based maintenance strategy under imperfect repair. *Reliability Engineering & System Safety* 144(December): 234–43. doi:10.1016/J.RESS.2015. Elsevier.

11

The Emergence of Industrial and Systems Engineering Principles and Practices in Disaster Management

Andrea M. Jackman

IBM Corporation

Mario G. Beruvides

Texas Tech University

CONTENTS

11.1 Introduction

Most adults, regardless of age, have a defining disaster in their lives where they remember with exceptional clarity their circumstances when hearing the news. In our current age, these events seem to range from the Kennedy Assassination in 1963, to the eruption of Mount St. Helens in 1980 or the Challenger Explosion of 1986, up to the terrorist attacks of September 11, 2001 or Hurricane Katrina in 2005. It is not uncommon, when discussing such moments in history, to ask "where were you when you heard…?"

In many ways, this novel focus on reaction is reflected across the emerging industry of emergency management. It is assumed, for example, that the assassination of another president is extremely unlikely due to fundamental changes in the approach to security tactics that came about as a result of President Kennedy's death. People saw a vulnerability they hadn't seen before and reacted appropriately to prevent future tragedy. Natural disasters are no different from man-made terrorism events in that regard—changes to emergency management processes, updated building codes in areas prone to specific disasters, and increased public awareness are all natural and reasonable reactions to catastrophic events.

Yet despite the complexities of these reactions and their crucial function in life-saving situations, emergency management is still not broadly considered as a discipline or industry and continues to be defined by reactionary principles instead of forward-thinking, proactive approaches to disaster mitigation or prevention. The agencies responsible for stepping in to prevent loss of life or damage to property are small relative to their task; the concept of civilians handling disasters rather than the military is still relatively recent. The emergency management workforce has been demonstrated in academic literature to be understaffed for the task of handling a catastrophic event, extremely diverse in skill sets not necessarily related to disaster or management science, and favoring emergency management positions as temporary or transitory, either between careers in other fields or toward retirement.

The principles of industrial engineering can provide much needed maturity to this industry still in its early stages. Recent developments in technology—where once emergency managers could operate in relative isolation, their every move is now tracked and reported by social media and the 24-h news cycle—may soon leave emergency managers with no choice but to begin focusing more on the "management" side of emergency management. This chapter explores the history of emergency management as an industry and discusses the many facets of industrial and systems engineering which would provide beneficial process maturity to the industry of disaster management.

11.2 Brief History and Overview of Emergency Management as an Industry

11.2.1 Transition from Military to Civilian Activity during the 1970s

When looking at the history of emergency management, especially in the United States, it is important to remember that even the terms "emergency management" or "disaster relief" are recent developments of the late 20th century. Prior to World War II, the concept of government-led disaster management was almost unheard of. Natural disasters were handled by individuals or their local organizations within the community; churches or social groups could provide some assistance in cases of extreme damage or loss. Governmental assistance was limited to special acts of Congress, although there was no formal process for qualifying for or receiving such help. In the early 20th century, the United States saw two major disasters which received national attention—the Galveston Hurricane of 1900 and the San Francisco Earthquake of 1906. Following the events in San Francisco, the federal government did exercise their option to quickly fund a relief effort, although it came several days after major donations were provided by private financiers and nongovernmental relief organizations. The federal relief package was considered far too small to provide meaningful relief and recovery without other sources providing assistance, although it demonstrated national awareness of the severity of the crisis in San Francisco and at least a token willingness to help from Washington DC. The federal response to San Francisco contrasted greatly with that for the Galveston Hurricane only 6 years prior. Damage was so severe that it took several days for messages to reach Washington, at which point they were discarded as too extreme to be plausible. Surely Galveston was exaggerating! Neither Congress nor then-President William McKinley took further action, and nearly all relief efforts to Galveston were provided by social or church groups, including the newly formed American Red Cross. Cleanup in some instances was also provided by Houston residents, sometimes forced at gunpoint to help out. In today's society, such negligence from the federal government would be unheard of, but it speaks to the political mindset and expectations of the early 20th century.

A second factor in viewing disasters as local responsibility was the cultural mindset that disasters and war were synonymous. Since the time of the first colonies through the end of the Civil War, citizens of the eastern United States were somewhat accustomed to have troops on the march and accepting both collateral and deliberate damage to land and property due to war. Natural disasters were not viewed much differently—just another form of potential harm that might come to a homestead. Occasionally, the government could be relied upon for assistance, but a safer option was to conduct your own rebuilding efforts, only calling upon neighbors or social organizations if in dire need.

The war-driven and militaristic disaster-as-attacker mind-set continues to this day, although perhaps not as starkly. Relief efforts during the World Wars of the 20th century were almost entirely devoted to postwar cleanup, planned and carried out by the military. Herbert Hoover, prior to becoming president in 1929, made a name for himself as a "Great Humanitarian" during post-World War I relief efforts. He was later dispatched as then-Secretary of Commerce to Mississippi during the Great Flood of 1927 for the same purpose. For the most part however, emergency management for domestic and peacetime disasters continued to be handled by the military exclusively through the Cold War. During the 1950s and 1960s, disaster sociologists began turning their attention to "civilian" applications, including the "Duck and Cover" campaign, and the "reduction and control and panic reactions" among the nonmilitary population (see Quarantelli, 1987). Small departments began to emerge within the defense agencies, as predecessors to Federal Emergency Management Agency (FEMA), with specific focus on the civilian response to a major catastrophic event. Finally in the 1970s, with the threat of external attack from nonnatural sources winding down, emergency management began to take on the look and feel of a professional discipline.

11.2.2 Four Phases of Emergency Management

The 1970s saw significant change for emergency management. In 1974, the Disaster Relief Act was passed by Congress to formalize the somewhat haphazard process of obtaining federal assistance following a major disaster—the same process that had been operating almost unchanged since the time of George Washington. The Disaster Relief Act of 1974 would lay the groundwork for future legislation and formalized processes for responding to disasters. In that same year, the National Governor's Association defined four phases of emergency management, which are still commonly used today among industry practitioners to categorize activities within the field. See Figure 11.1 for one summary of the four-phase model, noting the phases are considered to take place at all levels of government from local up to federal and be both cyclical and overlapping.

Around the same time as the development of the four-phase model and even a few years prior, academic research centers began to open with special focus on the study of disasters outside of the military. Centers at Ohio State University, the University of Delaware, and the University of Colorado specifically opened centers as extensions of other social science departments, with other universities later following suit in hazard-specific engineering and physical science for natural hazards. Because of this segregation among emergency management researchers according to their own "parent" disciplines or departments, literature reviews in emergency or disaster management can be extremely challenging. A diligent researcher must search extensively through journals covering many topics, ranging from highly technical engineering studies to more qualitative social work

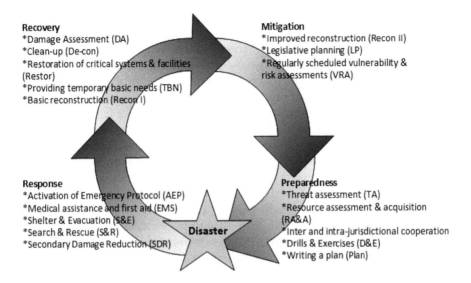

Recovery
*Damage Assessment (DA)
*Clean-up (De-con)
*Restoration of critical systems & facilities
(Restor)
*Providing temporary basic needs (TBN)
*Basic reconstruction (Recon I)

Mitigation
*Improved reconstruction (Recon II)
*Legislative planning (LP)
*Regularly scheduled vulnerability &
risk assessments (VRA)

Response
*Activation of Emergency Protocol (AEP)
*Medical assistance and first aid (EMS)
*Shelter & Evacuation (S&E)
*Search & Rescue (S&R)
*Secondary Damage Reduction (SDR)

Disaster

Preparedness
*Threat assessment (TA)
*Resource assessment & acquisition
(RA&A)
*Inter and intra-jurisdictional cooperation
*Drills & Exercises (D&E)
*Writing a plan (Plan)

FIGURE 11.1
The four phases of emergency management (Jackman et al., 2017).

case studies. Similarly, without an established academic discipline in emergency management, seminal papers and noted researchers are less visible. Sources are cited within this chapter where necessary; however, a more comprehensive list of source material and recommended reading is provided at the end of this chapter.

11.2.3 A Changing Political Environment

In 1988, Congress passed the Stafford Act, which provided clarity and refinement to the original process laid out in the Disaster Relief Act in 1974. While these two acts certainly provided clarity to government's role in disaster response, the other phases of emergency management (preparedness, recovery, and mitigation) were largely overlooked despite emergency managers spending most of their time outside of active response. The Disaster Mitigation Act (DMA) of 2000 was the first legislation to speak directly to mitigation and other preparedness aspects of emergency management, but before the requirements of DMA 2000 could take effect, the terrorist attacks of September 11, 2001 prompted a major shift among both policymakers and practitioners back toward man-made, warlike acts of terrorism. Numerous useful policies such as the National Incident Management System (NIMS) and subsequent Incident Command System (ICS) were created as a result, but attention was quickly refocused on natural disasters when Hurricane Katrina struck the Gulf Coast in 2005.

Hurricane Katrina represented a new era in emergency management in several ways. The massive storm occurred at a time when new technology in communication was just beginning to see widespread use. Cable news network broadcasting 24/7, mobile phones, and the Internet provided entirely new opportunities—some not even available for the September 11 attacks only a few years prior—for the world to watch a disaster unfold in real time. Without warning, emergency managers at all levels of government were thrust into the spotlight, following processes and procedures that were either too new to have been practiced or dated back several decades. What was supposed to be a clear-cut process for requesting aid from the federal government as laid out in the Stafford Act turned out to be both confusing and highly political (see Figure 11.2). The federal government took swift action in passing the Post-Katrina Emergency Management Reform Act in 2006. The bill had a patchwork feel, covering only specific areas of emergency management that seemed to have performed especially poorly during the hurricane, such as pet-friendly evacuations.

In addition to the unexpected yet thorough public attention, cultural and political forces contended that natural phenomenon on the scale of Hurricane Katrina were about to become normal, annual events due to changing climate conditions. Another response like the one to Katrina simply would not do, and emergency managers suddenly found themselves under intense scrutiny that has continued to this day.

11.3 Cultural Barriers to Quality Measures and Process Control

The public outcry following Hurricane Katrina was in many ways a reaction to a field or profession that was not well established, either publicly or internally among practitioners. Despite great strides by Congress and some academic centers in moving emergency management out of the military arena, the field was still operating largely as an afterthought to much larger government concerns such as healthcare, foreign affairs, and education. When media scrutiny arrived in the wake of Katrina, many emergency managers couldn't even say for sure if they were doing a good job or not. No one had ever defined what "good emergency management" was, especially in the new era of real-time disaster coverage.

11.3.1 Workforce Composition

One of the biggest ongoing challenges in professionalizing the field of emergency management is the prevailing workforce culture. As documented repeatedly in the literature (see References), emergency managers are typically retired military, police, firefighters, or similar first responders who may be reaching the end of their careers and are looking for a part-time position. These are typically job roles that highly value the ability to think rationally

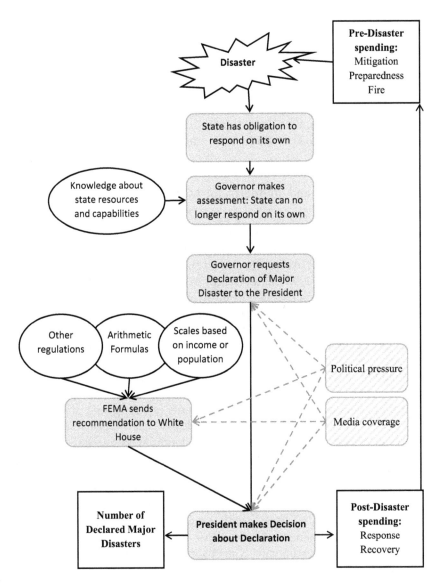

FIGURE 11.2
Overview of the disaster declaration process according to the Stafford Act of 1988 (Renken, 2016).

and act appropriately under life-threatening conditions. By their own admission (see Stuart Black et al.), emergency managers—when hiring—do not believe that someone without this type of experience is suitable for the job. Thinking back to Figure 11.1, it is easy to note however that only a small percentage of an emergency manager's job is spent in active response situations. The majority of his or her time will be spent in mitigation or preparedness and possibly some long-term recovery if a disaster occurred not long ago. Also by their own admission, emergency managers typically are not good at planning and preparedness activities and willingly rank themselves as performing poorly at such tasks (see Lindell, 1994). Several of these studies further cite political turf wars as major barriers to success at all levels of emergency management. Ingrained cultural biases between branches of the military or departments in local government, and against elected officials, can all contribute to an emergency manager's self-reported challenges.

11.3.2 Certifications and Standards for Measuring Professionals

Despite—or possibly because of—these ongoing struggles within the workforce, a small number of professional organizations attempted to develop certification programs with the field. The International Association of Emergency Managers (IAEM) developed the Certified Emergency Manager (CEM) certification in 1993 in an attempt to professionalize their workforce. The CEM requires demonstrated hours worked in all phases of emergency management and passing a written exam. However, there is some debate over the effectiveness of the CEM, as it also requires at the minimum a bachelor's degree in emergency management or equivalent field. For the reasons discussed in Section 11.3.1, some emergency managers feel a degree has little or no value in determining competency in a field where practical responder experience is valued above all else.

A similar certification exists for departments of emergency management, either at the state, county, or municipal level. The National Emergency Management Association (NEMA) pioneered a rigorous self-assessment and certification program for these departments. In recent years however, the popularity of the NEMA certification has declined. Critics of the program found the benefits of a NEMA certification too minimal to be worth the significant time and cost investment in obtaining it. In an informal survey conducted by the authors in 2016, it was found that fewer than 50% of state emergency management departments held an active NEMA certification. NEMA as an organization now primarily administers the Emergency Management Assistance Compact (EMAC) program, which establishes assistance agreements between states. This is a critical function in reducing waste during a response, yet still does not advance the definition of a good or professional emergency management operation.

It is no surprise, based on the cultural challenges listed above, that emergency managers have recently come under tight scrutiny from the public and media. Current technology and the ever-increasing complexity of society now place significant demands on emergency managers—a good emergency manager will not only have experience making rational decisions in life-threatening situations but must also demonstrate equal proficiency in mitigation and planning, social media and technology, politics, public relations, and the tightly coupled and interdependent systems that support modern society such as power or water distribution and waste removal. Without widely accepted quality measures or professional standards, anyone with an Internet connection or access to cable news can watch live coverage of a disaster unfolding and pass judgment on the performance of those in charge. When quality measures are applied in an academic setting, emergency management activities widely held to be inefficient may actually prove effective and vice versa.

11.4 Recommended Research and Application for Industrial and Systems Engineering Practices in Disaster Management

Emergency management as an industry has the potential to benefit greatly from the principles of industrial and systems engineering, although ultimately should strive to be its own academic discipline. The management of disasters has been researched from a variety of perspectives in the academic community, though always applying principles to specific components of emergency management through the lens of another, more established "field." Examples of this include the economic impacts of disasters (economics), shelter and evacuation case studies (sociology), or updated building design for specific hazards (civil engineering). While valuable, these studies inevitably disappear somewhat into their respective disciplines, rarely acknowledging emergency management as a discipline in and of itself. This is reflected across all of academics, not just research, with few schools offering a major or a department of emergency management. In many other fields, conferences are attended by both academics and practitioners, such that academics can present new research and practitioners can learn about the latest findings and potentially put them to use in the field. For any number of reasons, this is much less so in emergency management. The IAEM Annual Conference specifically solicits nonacademic speakers and presentations. Perhaps the case against such an offering was demonstrated by the lack of enthusiasm for the bachelor's degree by CEM critics, but it is clear from current practices and a rapidly changing societal landscape that something must be done to advance the field toward further professionalization and a better understanding overall of what constitutes "good" emergency management.

Industrial and systems engineering academics should consider and analyze if the topics in the following sections are suitable for inclusion in future IISE curricula and research.

11.4.1 The Application of Existing Process Management Principles to Emergency Management

Some topics are addressed in more detail below, but in general, the practices and industries described in this chapter would see tremendous benefit from an increased focus on the management aspect emergency management. Industrial and systems engineering can offer exactly the tools needed. A few areas ripe for research by industrial and systems engineers include the following:

- Performance measures in emergency management
- Quality and quality management measures in emergency management
- Planning challenges related to high-frequency events or high-impact/low-frequency catastrophic events
- Safety issues for a diverse yet part-time workforce
- Efficiency and effectiveness measures for both existing and proposed best practices
- Supply chain management, including inventory control and the application of systems dynamics

11.4.2 Quantitative Measurement of Policy Implementation

Perhaps the most immediate area where industrial or systems engineering would be application to disaster science is the measurement of success for processes and policies already in place. Major legislation was passed in 1974 (Disaster Relief Act), 1988 (the Stafford Act), 2000 (Disaster Mitigation Act), and 2006 (PKEMRA), yet without an academic support system—or even a commonly recognized think tank or talking head—no one seems to be able to answer the questions of whether these policies were good or bad for managing emergencies. How is the Stafford Act still being applied almost 30 years after its inception? As discussed earlier, Hurricane Katrina exposed major challenges in implementing the Stafford Act process for requesting aid. NIMS and ICS, as a result of September 11, 2001, are widely viewed as good guidelines for emergency managers, but who has measured their effectiveness? Was PKEMRA successfully implemented or are there issues still unaddressed from hurricane responses in the 2000s? Have any of the directed activities proven themselves to be new best practices or obsolete given changes in technology?

11.4.3 Workforce Development and Training

Another major challenge faced by emergency managers is how to staff a department. In the response mindset, all emergency managers would have significant response backgrounds. If you haven't pulled someone from a burning building, are you worthy of consideration for a position? An additional dilemma then arises—if responders are preferred, what should they be doing during the 90% (or more) of their time when there are no active disasters to respond to? In fact, the legislation mentioned in the previous section has already mandated how "downtime" should be spent, especially the DMA of 2000 which asks local-level responders to actively identify and remove vulnerabilities in their communities. This is a strategy not only asking responders to plan but plan to diminish their own role if proper preparations are made and a smaller response required as a result.

Similar to bridging the gap between research and practice, emergency managers need to continue moving toward a singular, professional workforce. While response backgrounds are still a critical need in response activities, the other four phases are underrepresented in terms of dedicated professionals. As an industry and career field, we must move away from transitional retirement positions where workers have backgrounds in other fields. IAEM has made strides in this area with the CEM program, but the structure of government and education must look to introduce changes.

11.4.4 Development of Phase-Specific Maturity Models

As a result of PKEMRA, FEMA developed a successful Logistics Capability Assessment Tool (LCAT), with implementation in over 30 states. The LCAT is a multiday self-assessment, similar to the aforementioned NEMA certification, for state emergency management offices to determine their own strengths and weaknesses in the specific area of logistics. FEMA does not require use of the LCAT, nor is it tied to any federal funding opportunities. The results are not made public. The LCAT exists purely to help states measure themselves using a maturity model. Tools such as the LCAT can and should be developed for other areas of emergency management, outside of logistics. From a research perspective, they provide natural, built-in data collection vehicles in areas where data is otherwise unavailable. They also provide tremendous benefit to the self-assessing agency, allowing state logisticians to privately determine their strengths and weaknesses, while being able to use the highest level assessment choices as targets for future growth.

References

Jackman, A.M., Beruvides, M.G. & Nestler, G.S. (2017). *Disaster Policy and its Practice in the United States: A Brief History and Analysis*. Buchanan, NY: Momentum Press; ISBN: 978-1606506998.

Lindell, M.K. (1994). Are local emergency planning committees effective in developing community disaster preparedness? *International Journal of Mass Emergencies and Disasters* 12(2), 159–182.

Quarantelli, E.L. (1987). Disaster studies: An analysis of the social historical factors affecting the development of research in the area. *International Journal of Mass Emergencies and Disasters* 5(3), 285–310.

Renken, K. (2016). Economic effects of mitigation spending in emergency management in the United States of America from 2004 to 2014. A Dissertation in Industrial and Systems Engineering, Lubbock, TX.

Critical Web-Sites

10th anniversary News Outlet Reports on Media Errors on Katrina: /www.the-guardian.com/us-news/2015/aug/16/hurricane-katrina-new-orleans-looting-violence-misleading-reports.

Certified Emergency Manager (CEM) Certification Program for Individuals: www.iaem.com/page.cfm?p=certification/faqs#req.

Emergency Management Assistance Compact (EMAC): www.emacweb.org/.

Emergency Management Accreditation Program (EMAP): www.emap.org/index.php/what-is-emap/who-is-accredited.

FEMA—Main Web-site: http://www.fema.gov/.

FEMA Assessments and Recommendations for Public Assistance: www.fema.gov/public-assistance-local-state-tribal-and-non-profit.

FEMA Assessments and Recommendations for Individual Assistance: www.disasterassistance.gov/.

FEMA Emergency Management Institute NIMS and ICS Educational Materials: https://training.fema.gov/emiweb/is/icsresource/.

National Emergency Management Association (NEMA): www.nemaweb.org/.

New York Times Summary of the Katrina Problem: www.nytimes.com/2005/09/19/business/media/more-horrible-than-truth-news-reports.html?_r=0The Stafford Act of 1988- PDF: http://www.fema.gov/pdf/about/stafford_act.pdf.

NOLA.com Katrina Flooding Map: www.nola.com/katrina/index.ssf/2015/08/katrina_flooding_map.html.

Office of Inspector General, and faced criticism of FEMA on Katrina: www.oig.dhs.gov/index.php?option=com_content&view=article&id=25&.

PKEMRA—Post-Katrina Emergency Management Reform Act of 2006: www.congress.gov/bill/109th-congress/senate-bill/3721.

Positive Changes Made within FEMA as a Result of PKEMRA: https://beta.fema.gov/pdf/about/programs/legislative/testimony/2011/10_25_2011_five_years_later_assessment_of_pkemra.pdf

Public and Media Criticism of FEMA: www.cnn.com/2006/POLITICS/04/14/fema.ig/.

Report from 2005: https://fpc.state.gov/documents/organization/53688.pdf.

The Stafford Act of 1988—Amendments—PDF: www.gpo.gov/fdsys/pkg/STATUTE-102/pdf/STATUTE-102-Pg4689.pdf.

Suggested Further Reading

Asch, Steven M., Stoto, Michael, Mendes, Marc, Valdez, R. Burgiaga, Gallagher, Meghan E., Halverson, Paul, Lurie, Nicole (2005). A review of instruments assessing public health preparedness. *Public Health Reports*, 120(3), 532–542.

Blanchard, B. Wayne (2006). The FEMA Higher Education Project. Retrieved Feb 8, 2007 from The Emergency Management Institute. http://training.fema.gov/EMIWeb/edu.

Boin, Ajen (2005). Disaster research and future crises: Broadening the research agenda. *International Journal of Mass Emergencies and Disasters*, 23(3), 199–214.

Borden, Kevin A., Schmidtlein, Mathew C., Emrich, Christopher T., Piegorsch, Walter W., Cutter, Susan L. (2007). Vulnerability of U.S. cities to environmental hazards. *Journal of Homeland Security and Emergency Management*, 4(2). doi:10.2202/1547-7355.1279.

Chang, Stephanie E. (2003). Evaluating disaster mitigations: Methodology for urban infrastructure systems. *Natural Hazards Review*, 4(4), 186–196.

Daniels, R. Steven, Clark-Daniels, Carolyn L. (2002). Vulnerability reduction and political responsiveness: Explaining executive decisions in U.S. disaster policy during the ford and carter administrations. *International Journal of Mass Emergencies and Disasters*, 20(2), 225–253.

Disaster Mitigation Act of 2000, Pub. L. No. 106–390, 114 Stat. 1552 (2000).

Downton, Mary W. and Pielke Jr, Roger A. (2002). Discretion without accountability: Politics, flood, damage, and climate. *Natural Hazards Review*, 2(4), 157–166.

Federal Emergency Management Agency. (1999). *Hazard Mitigation Grant Program Brochure- Desk Reference (No. 345)*. Jessup, MD: Federal Emergency Management Agency.

Federal Emergency Management Agency. (2002). *How-To Guide for State and Local Mitigation Planning (No. 386)*. Jessup, MD: Federal Emergency Management Agency.

Federal Emergency Management Agency. (2007). Approved Multi-Hazard Mitigation Plans. Retrieved 9/1/08 from www.fema.gov/plan/mitplanning/applans.shtm.

Federal Emergency Management Agency. (2007). *Design Guide for Improving Critical Facility Safety from Flooding and High Winds (No. 543)*. Jessup, MD: Federal Emergency Management Agency.

Freeman, Paul K. (2004). Allocations of post-disaster reconstruction financing to housing. *Building Research & Information*, 32(5), 427–437.

Ganderton, Phillip T. (2004). Cost Benefit Analysis of Disaster Mitigation: A Review (Working Paper). http://gandini.unm.edu/research/Papers/BCA_MitFIN.pdf.

Ganderton, Phillip T. (2005). 'Benefit–Cost Analysis' of disaster mitigation: Application as a policy and decision-making tool. *Mitigation and Adaptation Strategies for Global Change*, 10(3), 445–465.

Garrett, Thomas A., Sobel, Russel S. (2002). The Political Economy of FEMA Disaster Payments (Working Paper No. 2002–012B). The Federal Reserve Bank of St. Louis.

Jackman, A.M. & Martinez, M. (2007). Challenges in GIS utilization for rural emergency managers. *Proceedings of the ESRI GIS Homeland Security Summit*, Denver, CO.

Jackman, A.M. & Beruvides, M.G. (2008). Challenges in managing emergency planning activities in local government. *Proceedings of the 29th American Society of Engineering Management National Conference*, West Point, NY.

Jackman, A.M. & Beruvides, M.G. (2008). Multi-jurisdictional cooperation in hazard mitigation planning. *Presented at the Annual Hazards & Disasters Researchers Meeting*, Boulder, CO.

Jackman, A.M. & Beruvides, M.G. (2008). Risk and benefit cost analysis: interdependency in hazard mitigation planning. *Presented at the Annual Hazards & Disasters Researchers Meeting*, Boulder, CO.

Jackman, A.M. & Beruvides, M.G. (2013). Hazard Mitigation Planning in the United States: Historical Perspectives, Cultural Influences, and Current Challenges. In *Approaches to Disaster Management - Examining the Implications of Hazards, Emergencies and Disasters*, John Tiefenbacher, ed., InTech Publishing. ISBN; 978-953-51-1093-4.

Jackman, A.M. & Beruvides, M.G. (2013). How much do hazard mitigation plans cost? An analysis of federal grant data. *Journal of Emergency Management*, 11(4), 271–279.

Jackman, A.M. & Beruvides, M.G. (2013). Local hazard mitigation plans: A preliminary estimation of national eligibility. *Journal of Emergency Management*, 11(2), 107–120.

Jackman, A.M. & Beruvides, M.G. (2013). Measuring the quality of emergency management. *Proceedings of the 34th American Society of Engineering Management National Conference*, Minneapolis, MN.

Jackman, A.M. & Beruvides, M.G. (2015). The variational effects of jurisdictional attributes on hazard mitigation planning costs. *Journal of Emergency Management*, 13(1), 53–60

Jackman, A.M., Beruvides, M.G. (2018). The emergence of industrial and systems engineering principles and practices in disaster management. In *Emerging Frontiers in Industrial and Systems Engineering: Success through Collaboration*, Nembhard et al., ed., Taylor & Francis, LLC; in press.

Jackman, A.M., Nestler, G.S., & Beruvides, M.G. (2016). The Disaster Mitigation Act of 2000: Implications for the Practice of Risk Reduction in American Local Government. *Disaster Risk Reduction at the Local Level: A 2015 Report on the Patterns of DRR Actions at the Local Level*. United Nations Office for Disaster Risk Reduction. Available at: http://www.unisdr.org/campaign/resilientcities/home/manage_private_pages

Jackman, A.M., Beruvides, M.G. & Nestler, G.S. (2017). *Disaster Policy and Its Practice in the United States: A Brief History and Analysis*. Momentum Press; ISBN: 978-1606506998.

Jackman, A.M., Beruvides, M.G., & Renken, K. (2019). Disaster mitigation: An analysis of compliance versus disaster declarations and spending. *Proceedings of the 2019 American Society of Engineering Management National Conference*, Philadelphia, PA, Submitted February 2019.

Lindell, Michael K. and Perry, Ronald W. (2003). Preparedness for emergency responses: Guidelines for the emergency planning process. *Disasters*, 27(4), 336–350.

Martinez, M. & Jackman, A.M. (2007). GIS as a tornado preparedness and recovery tool. *Proceedings of the ESRI GIS Homeland Security Summit*, Denver, CO.

Nelson, Christopher, Lurie, Nicole, Wasserman, Jeffrey (2007). Assessing public health emergency preparedness: Concepts, tools, and challenges. *Annual Review of Public Health*, 28, 1–18.

Pelling, Mark (2002). The macro-economic impact of disasters. *Progress in Development Studies*, 2(4), 283–305.

Petak, William J. (1985). Emergency management: A challenge for public administration. *Public Administration Review*, Special Issue, 3–7.

Renken, K., Jackman, A.M., & Beruvides, M.G. (2017). Economic effects in emergency management over the last decade. *Proceedings of the 2017 Industrial and Systems Engineering Conference*, Pittsburgh, PA.

Renken, K., Jackman, A.M., & Beruvides, M.G. (2018). Quantifying the relationship between pre-disaster mitigation spending and major disaster declarations for US states and territories. *Journal of Emergency Management*, Submitted December 2018.

Renken, K., Jackman, A.M., & Beruvides, M.G. (2019). Quantifying the relationship between pre- and post-disaster from a mitigation perspective. *Journal of Emergency Management*, Submitted February 2019. Accepted; in press

Rose, Adam (2004). Economic principles, issues, and research priorities in natural hazard loss estimation. In *Modeling the Spatial Economic Impacts of Natural Hazards*, Y. Okuyama and S. Chang (eds.), Springer, Heidelberg.

Rose, Adam (2006). Defining and measuring economic resilience to disasters. *Disaster Prevention and Management*, 13(4), 307–314.

Rose, Adam, et al. (2006). Benefit-Cost Analysis of FEMA Hazard Mitigation Grants. Working Paper: July 25, 2006.

Schilderman, Theo (2004). Adapting traditional shelter for disaster mitigation and reconstruction: Experiences with community-based approaches. *Building Research & Information*, 32(5), 414–426.

Spence, Robin (2004). Risk and regulations: Can improved government action reduce the impacts of natural disasters? *Building Research & Information*, 32(5), 391–402.

12

Very Small Entities: The Business Proposition

Angela D. Robinson
International Council on Systems Engineering

CONTENTS

12.1 Introduction

International Organization for Standardization (ISO) learned that its standards were perceived as "too complicated" for use by very small entities (VSEs). In 2005, the first meeting of ISO/IEC JTC1 SC7[1] Working Group (WG) 24 was held, and the work on ISO/IEC 29110 series began.

At the 2009, the International Council on Systems Engineering (INCOSE) International Workshop (IW), the Association Françoise pour l'information Scientifique (AFIS) (French INCOSE Chapter), and INCOSE established the Systems Engineering for very small and micro enterprises (VSMEs) WG.

This chapter will discuss the importance of VSEs as a value proposition, how to use ISO/IEC 29110 as a business proposition, and the basic tools provided by the ISO/IEC 29110 to get a product started. The operational framework of the ISO/IEC 29110 Basic Profile Group will be introduced, which includes project management (PM) process, system definition and realization (SR), product development process (PDP) for limited resources, as well as roles, activities, and tasks to develop a product with as few as five (5) people using nine (9) deployment packages (DPs).

Engaging very small business partners for Industrial and Systems Engineering projects using an ISO standard designed just for them is a new frontier with unlimited opportunities for women and minorities to remove barriers to entry into business sectors with an unusually high bar to entry.

12.1.1 Impact of ISO/IEC 29110

Many governments realize that small and medium-sized enterprises (SMEs) matter because they are the sources of new jobs that stimulate the economy and improve productivity. From this point on, VSEs will be addressed. VSEs are organizations of up to 25 people and are important for their contribution to employment, innovation, economic growth, and diversity. Creating new quality jobs means jobs in high productivity sectors that offer a path to improved economic conditions. One of the major challenges for low and middle-income countries is maintaining and growing a skilled workforce that offers a bright economic outlook. According to the 2013 World Development Report, 600 million jobs are needed worldwide over the next 15 years to keep employment rates at their current level (World Bank, 2012).

Governments, nongovernmental organizations (NGOs), and donor organizations spend large amounts of money for targeted programmers and broader polices to enhance employment and the creation of new firms. Much employment in emerging economy countries are in the VSE arena,

---

[1] International Organization for Standardization/International Electrotechnical Commission Joint Technical Committee 1/Sub Committee 7.

these firms are often targeted by such interventions. Typical interventions include the provision of finance and financial services, entrepreneurship training, business support services, wage subsidies, and measures that transform the business environment. Despite these efforts, not much is known about which of these interventions are effective or, more importantly, under which conditions interventions work. This is a topic of considerable relevance, principally owing to the observed growth effects of cross-border venturing and the demonstrated capacity of VSEs to drive economic development at national, regional, and global levels. Limited firm resources and international contacts as well as lack of requisite managerial knowledge about internationalization have remained critical constraints to VSE internationalization.

Systems are becoming larger and more complex, providing an opportunity for the VSE to become a supplier to systems engineering organizations where the basic target for this standard is the acquirers of large organizations that are using the ISO/IEC/IEEE 15288. The concepts applied in the ISO/IEC 29110 are not limited to large organizations who are acquirers, other targets are large organizations with a process that may be too big for a small project, organizations with no Systems Engineer (SE) capability that wish to pilot a program, organizations that need to refine or improve their current SE practices.

12.1.2 Value of VSEs

VSEs are a value proposition as they are creating products and services for the consumer in a global market economy. Many VSEs develop and maintain system and software components used in systems, either as independent products or incorporated into larger systems; therefore, the contribution and recognition of VSEs as suppliers of high quality products is required. The ISO/IEC TR 29110-5-1-2, Software engineering—Lifecycle profiles for Very Small Entities (VSEs) —Part 5-1-2: Management and engineering guide: Generic profile group: Basic profile has been translated into Spanish, French, and Portuguese indicating the need for the products and services offered by VSEs.

The need for VSEs to follow the ISO/IEC 29110 standard are found in business sectors requiring compliance to many ISO standards, the ISO/IEC/IEEE 15288, and the ancillary standards for industries that need compliance. Business opportunities are opening in new sectors, the most recent one was brought to the INCOSE VSE WG in the way of new space programs. A large aerospace company would like to engage subcontractors in work on some of the new commercialized space projects. Access to the VSE PDP allows for women and minorities to engage with a large industrial partner proving untold opportunities to prosper and grow a business that could reach and train currently unrecognized talent in undiscovered talent pools.

12.2 A Product Development Approach for ISO/IEC 29110

In discussing the barriers to entry of difficult markets, the small entrepreneurial organization are addressed that needs to meet some level of compliance to gain entry. This approach is considered an effective way to seed SE in large organizations that are reluctant or resistant to engaging in some of the basic activities that produce a more effective and efficient product process. This is also an approach that can be used by large organizations with well-established and mature SE processes and procedures, but the organization PDP is too large for a small product group that need a small process, and it is still compliant to the organizational PDP. In terms of the VSEs, as a special case of small business, there are many sectors of commerce that are embracing the idea of a standardized practice. This could open financial incentives where venture capitalists could find these "start-ups" as a new field of investment opportunities.

INCOSE has been working on a Generic Basic VSE Profile Guide that distills the 68-page content of the ISO/IEC TR 29110-5-6-2:2014(E) to about 15 pages of essential guidance. Inherent in following the guidance of the ISO/IEC29110 are metrics, best practices, and business process models based on the extraordinary success of very large programs over more than 30 years (Figure 12.1).

12.3 Getting a Product Started

The ISO/IEC29110 series of reports in conjunction with the INCOSE SE Handbook is the foundation for the INCOSE developed DPs. A guide is in

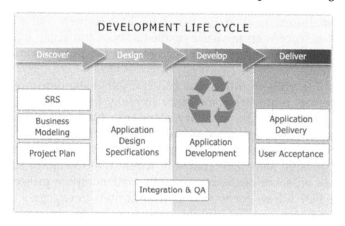

FIGURE 12.1
Product Development Lifecycle.

development for the standard's Generic: Basic Profile. The ISO/IEC29110 and DPs encompass the expertise and experience of more than 25 years of collective experience in the field of systems engineering.

The purpose of the INCOSE VSE WG is to encourage systems engineering concepts that have been adopted and adapted by most industries. In the context of VSE and small projects, these concepts can be tailored to improve product development efficiency and product quality, and contribute to standardization. The goal of the WG is to have systems engineering concepts available for small organizations who wish to work with many industries in various domains for the development of product and services.

The INCOSE VSE WG has a desire to improve and make product development within VSEs more efficient by using systems engineering concepts, standards, and proven practices. The WG further encourages guidance on process tailoring to lower barriers of entry to doing business as either a prime or subcontractor.

There are nine DPs, which comprise the system realization of product life cycle as well as the PM pieces, as illustrated in Figure 12.2. In the next four sections, the abbreviated process and activity models are presented, so the reader can get a sense of the ISO/IEC 29110 tool set that guides the VSE to not only success but quality and repeatable activities that allows the VSE to mature and grow.

The V-model represents a development process and demonstrates the relationships between each phase of the development life cycle and its associated phase of product integration testing shown in Figure 12.3.

12.3.1 The PM Process

The purpose of the PM process is to establish and carry out in a systematic way the *Tasks* of the system development project, which allows complying with the project's *Objectives* in the expected quality, time, and costs. There are

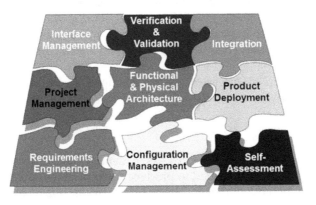

FIGURE 12.2
Systems Engineering Deployment Packages for the Basic Profile.

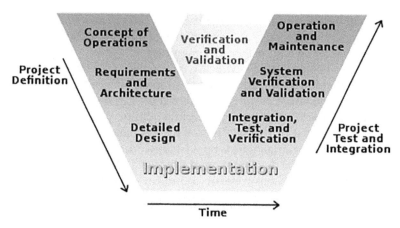

FIGURE 12.3
V-model.

eight PM objectives (PM.O), and these objectives may need document artifacts, archives, and evidence required by your regulated body. The following are the briefly stated objectives. While working through the PM section, it is important to note that each of the "plans" may be addressed by only one sentence included in the overall PM plan. The guidelines are to ensure that each of the requirements is at least considered during the project and planning process.

PM.O1. The *Project Plan* and the *Statement of Work* (SOW) indicate *Tasks* and *Resources* necessary to complete the work.

PM.O2. Progress of the project is monitored against the *Project Plan* and recorded.

PM.O3. *Change Requests* are addressed through their reception and analysis.

PM.O4. Review meetings with the work team and the stakeholders are held and tracked.

PM.O5. A *Risk Management Approach* is developed.

PM.O6. A Product Management Strategy is developed.

PM.O7. Quality assurance is performed to provide assurance that work products and processes comply with the *Project Plan* and *System Requirements Specifications* (SyRS).

PM.O8. A *Disposal Management Approach* is developed to end the existence of a system entity.

12.3.2 PM Activities

Defining the PM activities is the process of identifying and documenting the specific actions to be performed to produce the project deliverables. The key benefit of this process is that it decomposes work packages into scheduled activities that provide a basis for estimating, scheduling, executing, monitoring, and controlling the project work. This process is performed throughout the project.[2] This is a discipline that should be developed early in the product life cycle as it helps to assure good monitoring and control of the project, which is often the bane of many a small organization.

There are four basic PM activities, which are as follows:

PM.1 Project Planning: The project planning activity documents the planning details needed to manage the project.

PM.2 Project Plan: The project plan execution activity implements the documented plan on the project.

PM.3 Project Assessment and Control: The project assessment and control activity evaluates the performance of the plan against documented commitments.

PM.4 Project Closure: The project closure activity provides the project's documentation and products in accordance with contract requirements.

The PM process is laid out in a few simple constructs that are essential to the organization, tracking, and ultimately the delivery of a quality product using a highly collegial and collaborative team of up to 25 individuals.

12.3.3 SR Process

The purpose of the SR process is the systematic performance of the specification of system/system element, analysis, design, construction, integration, and verification/validation activities for new or modified systems according to the specified requirements. This is the technical portion of bringing the product to market or delivering to a customer. There are seven objectives in the system definition and realization (SR.O) process that allows the organization to reach a successful product launch. The following are the basic success objectives briefly stated.

SR.O1. *Tasks* of the activities are performed through the accomplishment of the current *Project Plan*.

SR.O2. System requirements are defined, analyzed for correctness and testability, and approved by the customer.

SR.O3. The system architectural design is developed and baselined.

SR.O4. System elements defined by the design are produced or acquired.

SR.O5. System elements are integrated.

SR.O6. A *System Configuration*, as agreed in the project plan.

SR.O7. Verification and validation *Tasks* of all required work products are performed using a defined criterion to achieve consistency among output and input products in each activity.

12.3.4 SR Activities

Finally, these are the activities which are required to be covered by the DPs. Each DP contains a detailed template for the PM and system realization activities as well as a mapping to the PM and SR processes.

---

[2] PMBOK Guide, Section 6.2 Define Activities.

The six essential SR activities are as follows:

SR.1 SR Initiation: The SR initiation activity ensures that the *Project Plan is* established.

SR.2 System Requirements Engineering: The system requirements engineering activity elicits and analyses the acquirer and other stakeholders' requirements, including legal and/or regulatory requirements.

SR.3 System Architectural Design: The system architectural activity transforms the system requirements to the system operational, behavioral, and physical architecture.

SR.4 System Construction: The system construction involves physical construction and/or software construction.

SR.5 System Integration, Verification, and Validation: The system integration, verification, and validation activity ensures that the integrated system elements satisfy the system requirements.

SR.6 Product Delivery: The product delivery activity provides the integrated system to the acquirer.

12.4 Using the Process to Build a Product

There was a need to demonstrate how DPs can be applied on a representative system application, the INCOSE VSE WG set out to identify a suitable case study. The search discovered the Autonomous Rover Case Study defined by the Eclipse Foundation[3] Polarsys Project9. The VSE WG partnered with the Polarsys project and developed a version of the Autonomous Rover Case Study that would serve as a project demonstration that was in the public domain.

The DP package structure contains the technical description in Section 1. Definitions are found in Section 2. Section 3 contains information about the relationship with the ISO/IEC 29110 standard, and Section 4 lists the description of activities, tasks, steps, roles, and products. Section 4 contains the description of processes, activities, tasks, steps, roles, and products, and describes how each package is utilized and is its core content. The remaining sections include templates, examples, checklist, and tools.

This chapter will utilize three DPs as the beginning of a product life cycle where it will detail the discover and design portion of the development life cycle as shown in Figure 12.1. These two stages are the most critical of the product life cycle because they determine the triple constraints of PM, time, cost, and scope. These phases utilize three DPs: PM, configuration management (CM), and requirements.

---

[3] www.eclipse.org/.

12.5 The PM DP

The process input for the PM piece is illustrated in Figure 12.4. Each DP has a prescribed organization, which is consistent throughout the packages. There is some variation in style because there were different authors, and

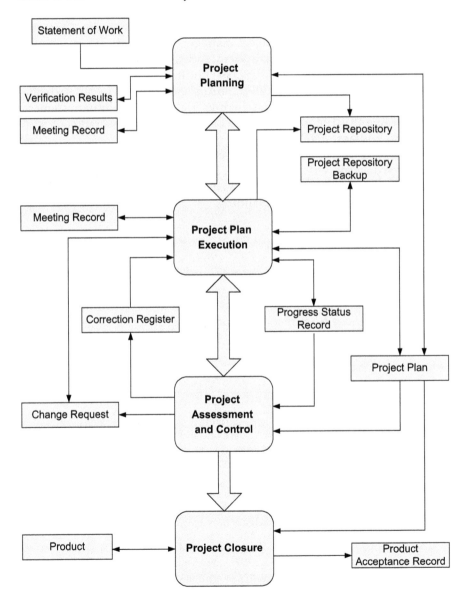

FIGURE 12.4
Overview of Project Management Process (ISO/IEC TR 29110-5-6-2:2014(E)).

these packages are in the process of being updated and peer reviewed to be standalone while being consistent in content and style. Each DP begins with why that package is important. For example, the PM DP indicates, "The purpose of the Project Management process is to establish and carry out, in a systematic way, the *Tasks* of the system development project, which allows complying with the project's *Objectives* in the expected quality, time and costs. Many systems fail not because there is no market, but because the cost of creating the system far outstrips any profit. Currently approximately half a million project managers worldwide are responsible for in the region of one million system and software projects each year, which produce products worth USD$600 billion. It is now accepted that many of these projects fail to fulfil acquirers' expectations or fail to deliver the system within budget and on schedule" (Jalote, 2002). Putnam suggests that about one-third of projects have cost and schedule overruns of more than 125% (Putnam, 1997).

The details of the project planning and project plan execution portions of this flow diagram will be examined. First, the roles are explained. Once defined, the next step is to define the activities and process flow for each phase being utilized for project planning.

12.5.1 Role Description for PM

This is a list of the roles, abbreviations, and competencies as defined in ISO/IEC 29110 5-6-2. The roles defined here are used consistently throughout the nine DPs. It is important for the team to understand the expected competency because team members may be required to be accountable to several roles.

	Role	Abbreviation	Competency
1.	Project manager	PJM	Leadership capability with experience making decisions, planning, personnel management, delegation and supervision, finances, and system development
2.	Technical leader	TL[a]	Knowledge and experience in the system development and maintenance
3.	Work team	WT	Knowledge and experience according to their roles on the project (e.g., SE, Engineer, Specialty Engineering)
4.	Acquirer	ACQ	Knowledge of the acquirer processes and ability to explain the acquirer requirements The acquirer (representative) must have the authority to approve the requirements and their changes The acquirer includes user representatives in order to ensure that the operational environment is addressed. Knowledge and experience in the application domain

(Continued)

	Role	Abbreviation	Competency
5.	Systems engineer	SYS	Knowledge and experience eliciting, specifying, and analyzing the requirements Knowledge in designing user interfaces and ergonomic criteria Knowledge of the revision techniques Knowledge of the requirements authoring Knowledge of the business domain Experience on system development, integration, operation, and maintenance Experience on the system development and maintenance
6.	Designer	DES	Knowledge and experience in the architecture design Knowledge of the revision techniques Knowledge and experience in the planning and performance of integration tests Knowledge of the editing techniques Experience on the system development and maintenance

[a] This role is not called out in the role section of the standard, but it is mentioned, and this process is designed to be tailored to the project.

12.5.2 Project Planning Process

The following activities are defined briefly in Section 3.2; here is the demonstration of activity PM.1.3. This activity is followed by the steps to create the work breakdown structure (WBS). This is a fundamental step in understanding all the project tasks that need to be completed. This is also the stage at which scheduling and resourcing is realized.

Activity: PM.1 Project Planning (PM.O1, PM.O5, PM.O6, PM.O7)

Project Planning Process PM.1.3

Tasks	Roles[a]
PM.1.1 Review the *SOW*.	PJM, TL
PM.1.2 Define with the acquirer the Delivery Instructions of each one of the deliverables specified in the *SOW*.	PJM, ACQ
PM.1.3 Identify the specific tasks to be performed in order to produce the deliverables and their system components identified in the *SOW*.	PJM, TL
PM.1.4 Establish the *Estimated Duration* to perform each task.	PJM, TL
PM.1.5 Identify and document the resources (i.e., human, material, equipment, and tools), including the required training of the work team to perform the project.	PJM, TL
PM.1.6 Establish the *Composition of Work Team* assigning roles and responsibilities according to the Resources.	PJM, TL

(Continued)

Tasks	Roles[a]
PM.1.7 Assign estimated start and completion dates to each one of the tasks in order to create the *Schedule of the Project Tasks* taking into account the assigned resources, sequence, and dependency of the tasks.	PJM, TL
PM.1.8 Calculate and document the project *Estimated Effort and Cost*.	PJM
PM.1.9 Identify and document the risks that may affect the project.	PJM, TL
PM.1.10 Document the CM strategy in the project plan.	PJM, TL
PM.1.11 Generate the *Project Plan* or update it. Furthermore, the *Project Plan* can be updated due to the *Change Request* made by the acquirer or arising from the project.	PJM
PM.1.12 Include product description, scope, objectives, and deliverables in the *Project Plan*.	PJM, TL
PM.1.13 Verification of the *Project Plan*. Verify that all *Project Plan* elements are viable and consistent.	PJM, TL
PM.1.14 Validation of the *Project Plan*. Validate that the *Project Plan* elements definition match with the *SOW*.	PJM, ACQ
PM.1.15 Establish or prepare the project repository using the *Version Control Strategy*.	PJM, TL

[a] Roles are defined in ISO/IEC 29110 Part 5-1.

Objectives	The primary objective of the project planning process is to produce and communicate effective and workable project plans. This process determines the scope of the PM and technical activities, identifies process outputs, project tasks and deliverables, and establishes schedules for project task conduct, including achievement criteria and required resources to accomplish project tasks.
Rationale	Whatever the size of the project, good planning is essential if it is to succeed. Effective system PM depends on thoroughly planning the progress of a project. A plan formulated at the start of a project should act as a driver for the project. The initial plan should be the best possible plan given the available information. It should evolve as the project progresses and better information becomes available.
Roles	Project manager
	Analyst
	Acquirer
Artifacts	Project plan
	Project description
Steps	1. Identify products and activities
	2. Create a WBS
	3. Estimate resources, effort, and duration
	4. Create a schedule

(Continued)

Step	*Step 1. Identify products and activities.*
description	During this stage, the project manager identifies all the products, tasks, and activities that need to be completed before the project can be finished. It may be necessary for the project manager to liaise with the acquirer and the analyst to fully understand the objectives of the project and to break down each one into its constituent parts.
	Step 2. Create a WBS.
	The WBS aims to identify all of the projects tasks that need to be completed and organizes them in a hierarchal format, where smaller subtasks contribute to the completion of a larger task at a higher level.
	A typical WBS would consist of the following:
	Project
	Task
	Subtask
	Work package
	Effort
	Once the WBS is complete, project milestones (key deliverables) can be identified and may be used for project tracking.
	Tips: Many system packages such as MS Project can structure WBS information and automatically generate useful graphical representations.
	Step 3. Estimate resources, effort, and duration.
	For each task in the WBS, the effort and duration should be estimated and the overall resources required to complete the project calculated.
	Typically, a "bottom-up" approach is used to estimate the effort required for each task in the WBS in terms of person hours or person days.
	In order to create a schedule of tasks and estimate the total project budget, it is necessary to estimate the resources (e.g., people, equipment, services) required to complete each task.
	Step 4. Create a schedule.
	Tasks should be organized into a coherent sequence, including parallel activities and mapped against time and resources, to produce a schedule of tasks to be completed by individuals during the lifetime of the project.
	Communications
	Confirm all previous steps are fully communicated to all parties by
	• Distributing all output reports to all parties, in particular those that must do the tasks or be influenced by the outcome of the tasks or schedules.
	• Holding a meeting(s) with all participants to ensure any queries are raised and actioned or assigned for action.
	Tips:
	When scheduling task execution ensure the percentage of time each participant has to allocate to the project is taken into account; thus, if the task should take 3 days full time and the task owner can only allocate 1 day a week then the task duration will take 15 days elapsed time.
	Many system packages such as MS Project can assist with capturing data about tasks in a WBS and generating activity network diagrams and Gantt charts.

12.5.3 Project Plan Execution

The activities for the project plan execution are indicated in the PM.2 process. The project plan execution exercise is where the team begins to determine the capability of the planning process. Identifying the WBSs and successful execution is managed and monitored through this process.

Activity: PM.2 Project Plan Execution (PM.O2, PM.O3, PM.O4, PM.O5, PM.O7)

Task	Roles
PM.2.1 Review the *Project Plan* and record actual data in *Progress Status Record*.	PM, TL, WT
PM.2.2 Analyze and evaluate the *Change Request* for cost, schedule, and technical impact and include the accepted changes in the *Project Plan*.	PM, TL
PM.2.3 Conduct revision meetings with the work team, review risk status, record agreements, and track them to closure.	PM, TL
PM.2.4 Conduct revision meetings with the acquirer, record agreements, and track them to closure.	PM, ACQ, TL, WT
PM.2.5 Perform CM.	PM, WT
PM.2.6 Perform *Project Repository* recovery using the *Project Repository Backup*, if necessary.	PM

Process: Project Management Process PM.2.1

Objectives	To implement the actual work tasks of the project in accordance with the project plan.
Rationale	Ideally, when the project plan has been agreed and communicated to all teams' members, work of the development of the product, which is the subject of the project, should commence.
Roles	Project manager
	Analyst
	Developer
	Acquirer
Artifacts	Project plan
	Project status record
	Change requests
Steps	1. Obtain agreement on project plan
	2. Record status
	3. Take corrective action
Step Description	*Step 1. Agreement on project plan.*
	Agreement must be reached between all the project managers and all members of the project team on the defined project parameters and targets as set out in the project plan. It may also be necessary to gain the agreement of the acquirer in terms of project duration and deliverables schedule.
	Step 2. Record status.
	The project manager should monitor and record the actual progress of the project against the planned progress. A record of actual project data should be maintained in a progress status record. To record status, a "traffic light system" could be used. The red/yellow/green traffic light approach is one used commonly in PM as it is a color theme everyone is familiar with. The colors are as follows:
	• Green—the task is "on target."
	• Yellow—the task is "not on target but recoverable."
	• Red—the task is "not on target and recoverable only with difficulty."

(Continued)

The typical contents of such a record are as follows:
- Status of actual tasks against planned tasks
- Status of actual results against established objectives/goals
- Status of actual resource allocation against planned resources
- Status of actual cost against budget estimates
- Status of actual time against planned schedule
- Status of actual risk against previously identified risk

Step 3. Corrective action.
When deviations between the project plan and actual project progress have been identified or the implementation of change requests agreed, corrective action will need to be taken to ensure than project continues according to revised plan.

The Rover PM plan document is based on the template construct found in the PM DP in the template section. Also included in this section are examples for WBS and sample project status template.

Project Plan—Sample Table of Contents

1. Introduction

   1.1 Project Overview
   A high-level overview of the project to include project objectives, list of the members of the team, and relationship (if any) to prior/existing projects.

   1.2 Project Deliverables
   A list of the items (e.g., documentation, code) to be delivered.

2. Project Organization

   2.1 Process Model
   A description of the process to be employed for the project.

   2.2 Project Responsibilities
   An identification of the roles and specific responsibilities to be adopted by each member of the project team.

   2.3 Change Control Procedures
   A description of how change will be handled.

   2.4 Configuration Management

3. Project Management Process

   3.1 Monitoring and Control Mechanisms
   A description of the methods to be used for monitoring progress and controlling the procedures that will be employed.

   3.2 Risk Management
   A description of main risks and risk mitigation strategy.

4. Work Packages, Schedule, and Budget

   4.1 Work Packages
   Description of the WBS and deliverables.

4.2 Resource
> The allocation of resources to the tasks.

4.3 Schedule
> Showing planned starting and finishing dates of each task listed and milestone.

4.4 Budget

Project Financial Plan

A description of how configuration will be implemented.

12.6 Configuration Management

CM is a systems engineering process for establishing and maintaining consistency of a product's performance, functional, and physical attributes with its requirements, design, and operational information throughout its life. This process has been tailored for the VSE in the ISO/IEC 20110. The standard includes CM as part of the PM process, PM.1, activities, and is a system realization objective, SR.O6. A *System Configuration*, as agreed in the project plan, and that includes the engineering artifacts are integrated, baselined, and stored at the *Project Repository*. Needs for changes to the *Product* are detected and related change requests are initiated.[4]

12.6.1 CM Roles

These roles are along the lines of the PM planning process, but executing on these roles is critical for this component of the product life cycle. Since CM is also an objective of system realization, the persons with these roles are integrated through the PDP in all phases.

	Role	Abbreviation	Competency
1.	Project manager	PM	Leadership capability with experience making decisions, planning, personnel management, delegation and supervision, finances, and system development
2.	Technical leader	TL	Knowledge and experience in the system development and maintenance
3.	Work team	WT	Knowledge and experience according to their roles on the project: SE, Engineer, Specialty Engineering, etc.

(Continued)

---

[4] ISO/IEC TR 29110-5-6-2:2014(E).

	Role	Abbreviation	Competency
4	Acquirer	ACQ	Knowledge of the acquirer processes and ability to explain the acquirer requirements
			The acquirer (representative) must have the authority to approve the requirements and their changes
			The acquirer includes user representatives in order to ensure that the operational environment is addressed
			Knowledge and experience in the application domain

12.6.2 Establish the CM Strategy (PM.O.3, PM.O.6)

Many large and mature organizations have very complex CM tools. The tracking of changes for design as well as planning documents is essential for a small team and is often overlooked. The CM strategy and execution will be elaborated in this section.

Activity: PM.1.10 Project Planning (PM.O1, PM.O5, PM.O6, PM.O7)
SR.O6. System Configuration, as agreed in the Project Plan

Objectives	Decide on an acceptable CM strategy consistent with the project plan, stakeholder expectations, and project expertise.
Rationale	The CM strategy should provide an appropriate level of configuration control for the project. The CM function controls the project baseline of documents, data, and product deliverables.
Roles	PM. The project manager has a leadership role in the CM strategy and process. PM ensures customer CM requirements are addressed by the project team.
	SE. The SE is responsible for CM. Depending on project size, the SE may perform CM functions or have a dedicated configuration manager.
Products	CM strategy (include in project plan)
	The CM strategy discusses how CM is performed on the project and the expected results. How is the project managing and controlling the project baseline? Who does the CM function and what reporting is required?
Artifacts	Project baseline configuration items. List the configuration items and make visible to the project team in the project repository.
	Revision control nomenclature and policy with authorizers.
Steps	1. Develop the CM strategy.
	2. Define CM tasks and integrate into project plan.
Step Description	*Step 1. Write the CM strategy.*
	Write the CM objectives as derived from project description, customer needs, and internal business requirements. The CM strategy can be a section or attachment to the project plan. Include a list of the known configuration item deliverables. This is a list from the customer or stakeholder. Description of the step, input/output, form used, etc.
	Step 2. Incorporate CM into the Project Schedule.
	Build CM deliverables and activities into the top-level schedule. Allow enough time to complete CM activities.

12.6.3 Conduct CM during Project Execution

Once the configuration strategy is established and codified in the project plan, process details on how to execute on that strategy follow. The first documents placed under CM are the internal documents generated as work product and artifacts as well as input documents from the acquirer and stakeholders. This is where a change management system is deployed, and CM methodology is established.

Activity: PM.1.10 Project Planning (PM.O1, PM.O5, PM.O6, PM.O7)
PM.2.5 Perform Configuration Management

Objectives	Execute CM on the project consistent with the strategy and schedule built in step 1.
Rationale	Once the CM strategy is developed, PM and SE ensure CM functions are accomplished to meet project needs. CM is critical to project success but need not be a complicated or involved process.
Roles	PM. The project manager assigns CM function to responsible team member(s). Depending on thw size of the VSE, the CM function may be done by the PM or the CSE.
	SE. The SE is responsible for CM. Depending on project size, the SE may perform CM functions or have a dedicated configuration manager.
Products	CM baseline
	The CM baseline is developed per the CM strategy. The CM baseline is approved by a project process and posted to the project repository.
Artifacts	CM baseline management process.
	Everyone on the project must know how the CM baseline is established and governed throughout the life cycle of the project.
Steps	1. Write the steps used on the project to develop the project CM baseline.
	2. Define how the CM baseline is changed or modified by members of the project team.
Step Description	*Step 1.* Develop the process of establishing the project baseline and how the project will manage the configuration-controlled baseline. This may be a simple list of the steps the project team uses to document and govern the project baseline.
	Step 2. Define and implement a change management process. This includes developing a change request and tracking of the change request through the process.

A sample CM plan is found in the template section of the CM DP.
Configuration Management Plan Outline

1. *Introduction*

   1.1 *Purpose*

   1.2 *Scope*

   1.3 *References*

   1.4 *Configuration Item Nomenclature*

   1.5 *Baselines*

12.7 Project Requirements Gathering

"Requirements Engineering provides a means of defining and communicating the capabilities, conditions and constraints which a system is required to satisfy. To paraphrase Lewis Caroll in Alice in Wonderland[5]:

"If you don't know where you're going, any road will get you there."

Requirements are considered to be the cornerstone of the Systems Engineering effort; therefore, modern Systems Engineering could not exist without first establishing quality requirements." In his systems engineering ROI paper, Eric Honour[6] found an optimal distribution of the Systems Engineering budget over the life cycle of a project, as shown in the following figure:

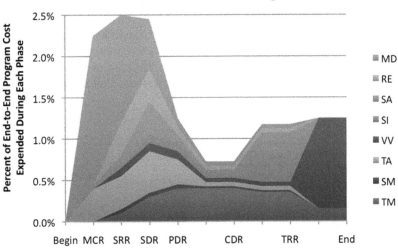

In the graph, requirements engineering activities[7] are represented by two areas:

---

[5] http://philosiblog.com/2011/07/13/if-you-dont-know-where-youre-going/.
[6] HONOUR Eric, "Systems engineering return on investment", Defence and Systems Institute, School of electrical and Information Engineering, University of South Australia, January 2013.
[7] The other activities shown in the graph are: Scope Management (SM); System Architecting (SA); System Integration (SI); Technical Analysis (TA); Technical Leadership/Management (TM); and Verification & Validation (VV).

- Mission/Purpose Definition (MD), by which stakeholder requirements are elicited and developed; and
- Requirements Engineering (RE), by which system and system elements requirements are elicited and developed.

Whereas the early emphasis of the requirements-related activities focused on technical (i.e., system and system element) requirements, recent research and experience indicate that the most important requirements (i.e., those driving system quality and system success) are stakeholders' requirements expressing their needs and expectations."[8]

The Rover demonstration case studies[9] are developed for the Dagu Electronics Rover5. This model was chosen for availability, low cost, and configurability.

The Rover project is used as a demonstration tool for the VSE process using the ISO/IEC standard and its adaptability to model-based systems engineering (MBSE). The INCOSE VSE WG demonstrates the ISO process first as a paper exercise. A future project under development is to turn the paper exercise into an MBSE exercise for teams that wish to develop proficiency in using modeling tools.

The Rover project basic bill of materials (BOMs) is presented as part of the gathering of the system functional requirements and the Rover model is pictured in Figure 12.5 and the Raspberry PI Controller Board is shown in Figure 12.6. This information along with the Autonomous Rover SyRS[10] V1_0 and Autonomous Rover StRS[11] V1_0[12] files are utilized to produce the Rover system design input requirements (SDIR) as well as elicit acquirer and other stakeholders' requirements.

The Rover Project Basic BOMs

Raspberry PI 3	Optional:
Pololu RPi hat motor controller 2756	Pi Camera
Infrared sensor	Pi Camera case
MCP 3008 analog/digital converter	Sonar sensor
Bread board	
Wires	
Battery with two USB ports	
One USB cable (for the RPi3)	
One USB cable to be modified to power the motor controller	

---

[8] Requirements_Engineering_Deployment_Package_VSME_V2.0.

[9] ISO/IEC 29110 Deployment Packages and Case Study for Systems Engineering: The "Not-So-Secret" Ingredients That Power the Standard.

[10] System Requirements Specification.

[11] Stakeholder Requirements Specification.

[12] GitHub Polarsys Autonomous Rover Systems Engineering Case Study Repository.

FIGURE 12.5
Dagu Electronics Rover5.

FIGURE 12.6
Raspberry PI 3.

12.7.1 Roles for Gathering Requirements

The requirements DP begins with making certain that the process is resourced with the appropriate personnel to achieve the required outcomes, documents, and artifacts for a successful product build. The standard indicates the product team can be up to 25 people; therefore, some people may fulfill multiple roles according to their skill set.

	Role	Abbreviation	Definition
1.	Systems engineer	SYS	Person in charge within the development team to gather, analyze, and manage the requirements related to the system to be developed
2.	Acquirer	ACQ	Person in charge within the acquirer side to transfer and validate requirements to the development team. It can be the acquirer or any representative
3.	Stakeholder	STK	An individual or organization that will be directly or indirectly affected by the use or deployment of the system.
4.	Designer	DES	Person(s) in charge of the design of the system or the system elements
5.	Reviewer		Person reviewing one or more system requirements engineering artifacts
6.	Project manager	PJM	Person in charge of managing the project (cost, schedule, tasks, contract, etc.)
7.	Work team	WT	All technical and specialists assigned to perform systems engineering, design engineering, or specialty engineering (safety, human factors, reliability and maintainability, etc.) activities on the project.

12.7.2 Project Plan Review

The first activity for the requirements process requires the team to review the plan because new roles are being executed by members that may not have been part of the planning process. The requirements engineering is defined by the *SR* process. The objectives described in Section 3.3 are utilized to begin work on the rover requirements. The activities are identified and the process for review of the plan is realized with the SR.1.1 process. This is an iterative process to keep the design and development on schedule or to adjust the plan.

Activity: SR.1 System Definition and Realization Initiation (SR.O1)

Tasks	Roles
SR.1.1 Review of the *Project Plan* by the work team to determine task assignment.	TL, WT
SR.1.2 Commitment to *Project Plan* by the work team and project manager.	PJM, WT
SR.1.3 Establish an implementation environment.	PJM, WT

SR.1.1—Review Project Plan with the Work Team members

Objectives	The objective of this activity is to review and revise, if necessary, the project plan with the work team as it relates to requirements engineering tasks, scope of the effort, the tailoring of tasks, as required, and task assignments.
Rationale	It is important that all members of the work team clearly understand the planned effort and operate "from the same sheet of music."

(Continued)

Roles	Team leader
	Work team
Artifacts	Revised project plan, as required
Steps	1. Assemble work team.
	2. Assign tasks.
	3. Distribute project plan.
	4. Hold requirements analysis kickoff meeting.
	5. Identify needed changes to project plan.
	6. Update and release updated project plan.
Step *Description*	*Step 1. Assemble work team.* The team leader identifies the member of the requirements engineering team and their respective roles based on the project plan. *Step 2. Assign tasks.* The team leader assigns each requirements engineering role to one or more team member, as required. *Step 3. Distribute project plan.* The team leader distributes the project plan to the work team for everyone to review the scope of the planned effort, whatever tailoring might have been planned and their respective role within the team. *Step 4. Hold requirements analysis kickoff meeting.* The team leader convenes and holds the requirements analysis kickoff to review the project plan, task assignments, and raise any issues for the work team as to needed revisions to the project plan. *Step 5. Review project plan changes.* The team leader reviews proposed changes and issues raised during the requirements analysis kickoff and analyses possible project plan changes. Step 6. *Update and release updated project plan.* The team leader incorporates needed changes to the project plan and releases the updated plan.

12.7.3 Elicit Acquirer and Other Stakeholders' Requirements and Analyze System Context

Based the Rover project,[13] the stakeholder and system requirements were analyzed to provide input to the requirements DP. The objectives, rationale, roles, and artifacts are described for teach of the activities of the SR.2 system requirements engineering process. The SR.2.1, "Elicit acquirer and other stakeholders requirements and analyze system context." and SR.2.4, "Elaborate System Requirements and Interfaces." will be examined in detail.

Detailed information for the task list for the SR.2 process is available in the ISO/IEC TR 29110-5-6-2:2014(E) report. Following the task list is the step-by-step instruction set to achieve the SR2.1 activity and produce the StRS for the demonstration project.

---

[13] https://polarsys.org/wiki/Systems_Engineering_the_Autonomous_Rover.

Activity: SR.2 System Requirements Engineering (SR.O2, SR.O6, SR.O7)

Tasks	Roles
SR2.1 Elicit acquirer and other stakeholders' requirements and analyze system context.	SYS, ACQ, STK
SR2.2 Verify the *Stakeholders Requirements Specifications* with PJM.	PJM, WT
SR2.3 Validate the *Stakeholders Requirements Specifications* with the acquirer and other stakeholders.	PJM, SYS, ACQ, STK
SR2.4 Elaborate system requirements and interfaces.	SYS, DES
SR2.5 Elaborate *System Elements Requirements Specifications* and the *System Interfaces Specifications*.	SYS, DES
SR2.6 Verify and obtain work team agreement on the *System and System Elements Requirements Specifications*.	PJM, WT
SR2.7 Validate that *SyRSs* satisfies *Stakeholders Requirements Specifications*.	SYS, DES
SR2.8 Define or update traceability between requirements.	PJM, TL
SR2.9 Establish or update the *IVV plan* and *IVV procedures* for the system verification and validation.	SYS, IVV

SR2.1 Elicit acquirer and other stakeholders requirements and analyze system context.

Each of the steps described in this process is designed to produce the stakeholders requirements specification (StRS) document. For VSEs, this is possibly the most critical part of the design process, it is during these discussions with the stakeholders that the imagineering takes place.

The following table outlines the process by which the stakeholder needs are captured, codified, and placed under CM.

Objectives	The main objective of this activity is to • Gain a comprehensive view of the acquirer and stakeholder needs. • Clearly define the scope of the project. • Elicit and analyze acquirer and stakeholder requirements source material.
Rationale	It is important to clearly define the project scope (boundaries), identify all sources of requirements, and identify key requirements of the future system with the stakeholders to avoid problems like forgotten key functionalities/characteristics or requirements creep.
Roles	SE Acquirer Stakeholders
Artifacts	StRS placed under configuration control
Steps	1. Identify stakeholders. 2. Collect information about the project and application domain. 3. Define project's scope. 4. Identify and capture stakeholders' source requirements. 5. Structure and prioritize stakeholders' requirements documents/materials.

(Continued)

Step	*Step 1. Identify stakeholders.*
Description	The SE, with the assistance of the project manager and the acquirer, identifies and establishes contact with the project's stakeholders. This can occur in a single meeting, a series of meeting or through individual, one-on-one meetings.

Example of stakeholders are as follows:

- Final users or final users representatives
- Community of users
- Acquirers or acquirers representatives
- Executive board
- Regulatory agencies (national, regional, local)
- Standardization agencies (international, national, regional)
- Employee unions
- Partners
- Testers
- Manufacturers
- Distributors
- Logistics specialists
- Maintenance, repair, and overhaul organizations
- Environmental protection organizations

Step 2. Collect information about the domain.

The SE captures the key concepts of the business domain of the stakeholders. The project manager, acquirer, and stakeholders assist the SE by providing all the information (existing documentation or explanation) that will facilitate this understanding.

Step 3. Review project's scope.

Review the project objectives and priorities as defined by the PM. Review system boundaries and context as identified by the F&PA DP.

Step 4. Identify and capture source requirements.

Having in mind key concepts related to the stakeholder business domain, the SE starts identifying requirements. None of the situations in projects are identical. In some cases, most of the requirements are already identified in a document (call for tender in case of fixed priced projects). However, in many cases, the requirements are implicit and not written. Some requirements can contradict themselves between stakeholders.

The SE identifies and lists the key requirements of the system to be built. During this step, the SE should not start detailing identified requirements. The main goal is to gain a comprehensive view of the needs.

The SE captures existing requirements source documents, placed under configuration control by the team leader or project manager. Each requirement/statement/paragraph may be assigned a project-unique identifier at this step to facilitate the creation of traceability links later on. The stakeholder source documents are typically not reformatted in any way, so as to facilitate communications with the acquirer and stakeholders as well the future updating of the document if changes are released.

12.7.3.1 Stakeholders Requirements Specification

The Rover project comes with user needs and intended uses that include scenarios which are used to synthesize the stakeholder experience and can be used to elicit the needs of the stakeholders. For these requirements, a unique identifier is used followed by a number. Establishing a system of assigning unique identifiers to requirements is needed for traceability, verification, validation, and tracking of requirements. For this exercise, the unique identifier STK.xx is used to identify the needs and intended uses, which is followed by a complete description of the requirement.

STK.01: The acquirer is a regional hazardous material emergency response unit.

Text: The acquirer is a regional hazardous material emergency response unit operating under the authority of the regional government.

STK.02: The hazardous material emergency response technician is trained to deploy, operate, and recover the autonomous rover.

Text: The hazardous material emergency response technician, also referred to as the response technician, is the sole person allowed and trained to deploy, operate, and recover the autonomous rover. Depending on the situation, the unit may dispatch one or two hazmat emergency response technician into a theater of operations.

STK.03: The emergency response unit has one maintenance technician on staff that will be qualified to perform any maintenance on the rover.

Text: The maintenance technician is capable of performing basic electronics soldering and assembly. The technician, however, is not qualified to perform high-reliability soldering or "clean room" fabrication/assembly tasks.

STK.04 Maintenance technician is not expected to reprogram the rover.

Text: The Maintenance technician will not be required nor is s/he expected, to be able to reprogram the autonomous rover beyond loading the operational program into the autonomous rover's memory.

STK.05: Maintenance technician's environment.

Text: The maintenance technician will not normally be dispatched to a theater of operations.

STK.06: The HazMat response team will always operate in support of a unit requesting their services.

Text: Other than for training exercises, the HazMat response team will always operate in support of a unit requesting their services. The unit can either be law enforcement, fire protection, emergency medical service, or armed forces.

STK.07: The HazMat emergency response unit will not provide "live" data.

Text: The HazMat emergency response unit will not provide "live" data nor will they transfer any of the data from the autonomous rover to the supported unit on site. If such a requirement exists, they will acquire the data through usual request for information submission administrative channels.

STK.08: Need to monitor autonomous rover sensor data or imagery "live" during a mission.

Text: A number of units have expressed a need to monitor autonomous rover sensor data or imagery "live" during a mission. This capability may be incorporated at a later time.

12.7.4 Elaborate System Requirements and Interfaces

The requirements gathering for the design functionality is part of the rover project requirements specification found in Section 6. This process determines the system requirements and any requirements not explicitly stated that may need to be derived.

Activity: SR.2 Specification and Agreement on the System Requirements (SR.O2, SR.O6, SR.O7)

SR2.4 Specification and Agreement on the System Requirements

Objectives	The objective of this step is to • Review the system boundary. • Capture the system requirements and system design constraints. • Capture the interface requirements between the system and its external environment. • Release a draft *SyRS for review.*
Rationale	The SE needs to establish a clear and complete definition of the system requirements.
Roles	SE Designers
Artifacts	System requirements database SyRS System prototype (optional)
Steps	1. Produce a prototype (optional). 2. Capture system requirements. 3. Prepare draft SyRS.
Step Description	*Step 1. Produce demonstrators, prototypes, analysis models, or simulations (optional).* The SE and designers can use demonstrators, prototypes, analysis models, or simulations to facilitate the comprehension and scoping of the stakeholder needs and system requirements (i.e., acquirer/stakeholder and development team side). A prototype may be used to implement only the subset of the intended system functionality, which represents the highest risk or is the most poorly understood. It must be understood that these mechanisms ARE NOT the intended solution. *Step 2. Capture system requirements.* The system engineer extracts system requirement statements from the source documentation, stakeholder interview notes, or as the result of the prototype production and evaluation. Typically, the source material/document paragraphs consist of many sentences and statements, not all of which are meant to be system requirements. The SE may use keywords such as "shall" and "must" in source document sentences to identify potential system requirements. Also, the SE sorts program/project requirements (i.e., the supplier shall…) from technical requirements (i.e., the system shall…) and retains only the technical requirements. Program/project requirements are passed on the project manager to action as required. If the source documents include a compliance matrix associated with a proposal document to the acquirer, the compliance matrix is used to convert each statement in the document referenced by the compliance matrix into fully compliant requirements (i.e., noncompliances are removed and partial compliances are applied so that only the "compliant" portion of the statement becomes a requirement).

(Continued)

The SE interacts with acquirer representatives, subject-matter experts, and stakeholders in order to clarify specialty requirements such as human factor engineering issues, whenever relevant (e.g., if a graphical user interface must be developed).

Step 3. Prepare draft SyRS.

The SE organizes the system requirements into a draft SyRS in accordance with the guidelines of ISO/IEC/IEEE 29148, Section 8.3.

Below is an example of the output of the requirements process that will be contained in the specification document and placed under CM. The requirement types have a unique identifier prefix appended by a number. For the functional requirements, REQID.xx is used. Similarly, performance requirements are identified by PREQID.xx. The naming convention is determined and agreed upon by the project team.

Functional Requirements

REQID	Requirement Text	Verification Method
REQID.01	The rover shall support two sensors from the following list: • IR sensor • Explosive gas sensor • Smoke sensor • Ammonia sensor • IR range finder • Acoustic (i.e., ultrasound) range finder • Radioactivity sensor • Water/liquid sensor • Light sensor • Carbon monoxide sensor	Inspection
REQID.02	The rover shall be able to deploy and retract a vertical boom at the end of which a sensor is mounted.	Demonstration
REQID.03	The rover shall explore a flat room (without holes) autonomously.	Demonstration
REQID.04	The rover shall produce a map of the explored environment.	Demonstration
REQID.05	The rover shall avoid crashing into obstacles.	Demonstration
REQID.06	The rover shall differentiate walls from inclined pathway.	Demonstration
REQID.07	The rover shall avoid falling on its backside.	Demonstration
REQID.08	During the mission the rover should adopt an avoidance and evasion strategy if it encounters a threat or danger.	Demonstration
REQID.09	Once the rover has completed the mapping and data collection mission, it is to return to the starting point where it was placed at the start of the mission.	Demonstration
REQID.10	It should be possible to abort a mission, at which time the rover will return to the point where it was placed at the start of the mission.	Demonstration
REQID.11	All collected sensor data and the area map shall be retained during power down.	Test
REQID.12	It shall be possible to download the map and sensor data from the rover using a cabled interface.	Test
REQID.13	It shall be possible to download the map and sensor data from the rover using the tablet controller.	Test
REQID.14	The rover shall differentiate walls from inclined pathway.	Demonstration

Performance Requirements

PREQID	Requirement Text	Verification Method
PREQID.01	The rover electrical power source shall provide an autonomy of at least 10 m.	Test
PREQID.02	The rover electrical power source shall provide an autonomy of at least 30 m.	Test

System Interface Requirement

SIREQ	Requirement Text	Verification Method
SIREQ.01	The only system interfaces for the rover should be the man–machine interfaces, including the following: • Basic controls for manual operations • Tablet controller either an Android or an iPad • An optional alternating current interface to recharge the Rover power source • The rover controller reprogramming/data download interface	Test

System Integration Requirements

SINTREQ	Requirement Text	Verification Method
SINTREQ.01	The rover shall be controllable through an Android application on an Android Tablet.	Demonstrate
SINTREQ.02	The rover shall be controllable through an IoS application on an iPad Tablet.	Test
SINTREQ.03	It should be possible to learn its operation in no more than a one training session.	Demonstrate
SINTREQ.04	The emergency response technician should become totally proficient with all operational aspects of the rover in no more than five training exercises.	Demonstrate
SINTREQ.05	A rover stowed in its utility case shall be transportable by a single person.	Demonstrate

System Maintainability Requirements

SMREQ	Requirement Text	Verification Method
SMREQ.01	It shall be possible for the emergency response technician to replace rover sensors and the batteries if they are not rechargeable.	Demonstration

The SyRS can be based on the template construct found in the requirements DP in the template section displayed in Table 12.1.

TABLE 12.1

SyRS Template Table of Content

1. Introduction	8. System modes and states
1.1 Purpose	9. Physical characteristics
1.2 Document conventions	9.1 Physical requirements
1.3 Intended audience	9.2 Adaptability requirements
1.4 Additional information	10. Environmental conditions
1.5 Contact information/SyRS team	11. System security
members	12. Information management
1.6 References	13. Policies and regulations
2. System overview	14. System life cycle sustainment
2.1 System context	15. Packaging, handling, shipping, and
2.2 System functions	transportation
2.3 User characteristics	16. Verification
3. Functional requirements	17. Assumptions and dependencies
4. Usability requirements	18. Other Requirements
5. Performance requirements	Appendix A: Terminology/Glossary/
6. System interfaces	Definitions list
7. System operations	Appendix B: Requirements traceability
7.1 Human system integration requirements	matrix
7.2 Maintainability	This matrix traces each requirement to
7.3 Reliability	its parent/source requirement or material.

Source: Adapted from ISO/IEC/IEEE 29148:2011, Section 9.4.

12.8 Conclusion

The INCOSE VSE WG, working with the ISO/IEC29110 series, exists to improve and make product development within VSEs more efficient by using systems engineering concepts, standards, and proven practices. This chapter is an introduction to methodology on tailored guidance to VSEs and how it may be applied in the context of their role as either a prime contractor or subcontractor. Fundamentally, it is the INCOSE VSE WG hope to achieve access for the VSE to the best of systems engineering techniques, methods, and processes for the betterment of the broad scope of domains around the globe.

References

INCOSE *Systems Engineering Handbook: A Guide for System Life Cycle Processes and Activities*, 4th Edition, Hoboken, NJ, Wiley, 2015, The International Council on Systems Engineering.

ISO/IEC 29110-1 Systems and Software Engineering—Lifecycle Profiles for Very Small Entities (VSEs)—Part 1: Overview.

ISO/IEC JTC 1 SC 7 Working Group 24 10th Anniversary Overview of accomplishments Prepared by Claude Y. Laporte, P. Eng., Ph.D. ISO/IEC 29110 Project Editor.

ISO/IEC TR 29110-5-6-2:2014(E) Systems and software engineering—Lifecycle profiles for Very Small Entities (VSEs)—Part 5-6-2: Systems engineering—Management and engineering guide: Generic profile group: Basic profile.

Jalote, P., *System Project Management in Practice*, Addison-Wesley, Boston, MA, 2002.

Laporte, C. Y., Alexandre, S., O'Connor, R., A Software Engineering Lifecycle Standard for Very Small Enterprises. In R. V. O'Connor et al. (Eds.): EuroSPI 2008, CCIS 16, pp. 129–141.

Project Management Institute PMBOK Guide, 6th Edition, Project Management Institute, Inc., Newtown Square, PA, 2017.

World Bank Development Report 2013, The World Bank, Washington, D.C., 2012. ISBN 978-0-8213-9620-9.

13

Countering Human Trafficking Using ISE/OR Techniques

Maria Mayorga, Laura Tateosian, German Velasquez,
and Reza Amindarbari
North Carolina State University

Sherrie Caltagirone
Global Emancipation Network

CONTENTS

13.1 Introduction

Human trafficking is the world's fastest growing criminal economy and the third largest income generator overall, generating $150 billion in revenue each year (ILO, 2017). Estimates suggest that 40 million people are enslaved around the world (ILO, 2017). Survey methodologies are inconsistent at best, rely on willing participants, and systematically underestimate the affected population. Counter-trafficking efforts have remained largely unchanged, and the ability to modernize the effort lies within a fractured stakeholder community. Law enforcement, government agencies, nonprofit organizations, technology companies, and commercial enterprises all have relevant data and skills that could drastically increase the number of victims identified and rescued as well provide evidence to secure traffickers' convictions.

Trafficking in all its forms represents a massive challenge to these counter-trafficking stakeholders. In the digital age, human trafficking has evolved into a complex system that leverages the power of technology to build supply chains, create sophisticated communication techniques, and protect criminals from scrutiny, particularly online. In parallel, these systems rely upon non-digital channels of communication, transportation, and recruitment. These channels have been continuously refined over centuries, and intentionally exploit the fractured flow of information and investigations. A coordinated, holistic approach that ultimately raises risk and cost to traffickers to unacceptable levels is critical. We submit that the key to this multipronged approach is uniting stakeholders and their datasets behind an industrial and systems engineering (ISE)-centered analytic framework to affect maximum disruption. Before discussing potential approaches, we briefly review the history of efforts to combat human trafficking, who is impacted, and the implications of the trade on society.

13.1.1 Fighting Human Trafficking: The Four Ps Paradigm

The means for fighting human trafficking can be framed as four types of actions, the four Ps: prevention, protection, prosecution, and partnership (Konrad et al., 2017). Each of these efforts requires resources; thus, analytical approaches can be useful to guide decision-making.

Prevention involves media campaigns to educate the public. Advertisements can be targeted at vulnerable segments to alert potential victims to the dangers of trafficking or it can be aimed at enabling others to recognize and report possible trafficking situations to authorities. Though social media has played a large part in facilitating illicit networks, raising awareness is one way in which social media can be used to fight the issue. Analytical models can help with optimizing advertising funding allocation, by estimating the reach of a campaign to target communities and determining when to modify advertisements because the message is becoming stale.

Protection refers to recognizing possible victims, interdicting trafficking networks, and providing shelter for rescued victims. Analytical tools have great potential to advance these first two fronts because of the trafficking industry's heavy reliance on social media. Online data can be collected and analyzed for indicators of trafficking networks. Then, optimization models can be applied to the networks to identify efficient usage for interdiction resources. Protection will be the focus of our detailed discussion to follow in Sections 13.3 and 13.4. *Prosecution* refers to the enforcement of anti-trafficking laws, the punishments upon conviction, and the policies surrounding human trafficking. The number of traffickers convicted lags greatly behind the number of victims identified, partially due to community stigmas associated with witnesses for testifying against perpetrators. Increased use of biometrics presents an opportunity for improvements in this area. The timeframe of prosecution is also important, as delays can reduce the detail and accuracy of victim testimonies. This situation can be improved with the use of queuing models.

The fourth P, *partnership*, refers to the need for organizations, including law enforcement, NGOs, and health care organizations, to share data in a way that protects privacy but enables investigators to "connect the dots." For example, many victims seek health care more than once while being trafficked, but the records are fragmented, so patterns are not recognized. Establishing a central repository with standardized data collection and login roles for privacy protection would improve sharing and facilitate more and faster discovery of trafficking activities.

13.1.2 Policy and Human Trafficking

The United Nations' protocol to prevent, suppress, and punish trafficking in persons (TIP) (UN Palermo Protocol, 2001) has been the most influential international agreement against human trafficking (Baer, 2015; Cho, 2015). It was a landmark achievement because for the first time, this protocol provided a definition for "human trafficking" with a baseline consensus, distinguished it from human smuggling and illegal migration (Gozdziak and Collett, 2005), and provided a means for identifying victims. This protocol also proposed the 3P (prevention, protection, and prosecution) classification of antihuman trafficking activities (later extended by the Department of State, to include a fourth P (partnership) as discussed in Section 13.1.1). However, according to Gozdziak and Collett (2005), the protocol excessively focused on criminalizing traffickers rather than protecting victims.

The United States signed the Protocol in 2000 and ratified it in 2005. The United States, however, passed its first Trafficking Victims Protection Act

(TVPA) in 2000, independent of the United Nations protocol. TVPA 2000 established methods for prosecuting traffickers, preventing human trafficking, and protecting victims (Polaris Project). In addition to setting severe punishments for human trafficking, and mandating restitution for the victims, it established the Office to Monitor and Combat Trafficking in Persons for preventing human trafficking. TVPA has been reauthorized in 2003, 2005, 2008, and 2013 with some modifications and expansion, such as a pilot program for sheltering minors who survived human trafficking, a pilot program for the treatment of trafficking victims abroad, a new system for collecting and reporting human trafficking data, and emergency response provisions within the State Department for disaster areas (Polaris Project).

13.1.3 Human Trafficking as a Public Health Issue

The catastrophic mental and physical health consequences of human trafficking are not limited to the victims and individuals that are directly involved but the welfare of an entire community. The poor mental and physical health of the victims of human trafficking bar them from active engagement in and contribution to their communities. The spread of communicable diseases through the unsanitary living and working conditions of the victims, including unsafe sex practices in the case of sex trafficking, put the entire society at risk. Human trafficking is a threat to the well-being of the entire community as it exacerbates inequalities that exist in our society by affecting the vulnerable populations the most (Chisolm-Straker and Stoklosa, 2017).

The magnitude of human trafficking and its direct and indirect impact on large numbers of people and multiple societal systems makes it a widespread public health issue. However, antihuman trafficking actions have so far been directed through law enforcement and the criminal justice system, which largely focuses on the well-being and actions of individuals. A public health approach to human trafficking, however, focuses on the well-being of communities rather than individuals. It allows us to understand the problem holistically and prevent it by addressing the root causes (Chisolm-Straker and Stoklosa, 2017).

13.1.4 Challenges in Countering Human Trafficking

Counter-human trafficking strategies have largely remained unchanged since the advent of the Internet. Two primary challenges make modernizing the effort difficult: a fragmented stakeholder community with siloed data and the failure to inject the right technologies that would enable scalable efforts despite the limited resources available. The technologies discussed in this chapter provide an opportunity to lessen both of these challenges.

In the following sections, we provide recommendations on how these challenges can be overcome by using ISE techniques. Interactions between perpetrators and victims of human trafficking and its causes and outcomes form a complex system. Operations research (OR) and ISE offer a framework for understanding complex problems. Analytics and techniques that OR/ISE offer not only help us to examine complex human trafficking operations and identify where and why these activities occur they also inform how to use resources more efficiently and predict future trends. Given its underground nature, reliable and quality data on human trafficking activities is scarce. OR and ISE techniques provide us with a means to extract information from large and even unstructured data—including images, communications on message boards, and online advertisements (Konrad et al., 2017).

In Section 13.2, we first review the fundamental concepts and components involved in human trafficking. In Section 13.3, we review geospatial and temporal data analytics that can help us improve understanding about human trafficking. Finally, in Section 13.4, we show the application of mathematical optimization techniques to efficient allocation of resources for tackling human trafficking. A conclusion is provided in Section 13.5.

13.2 Understanding Human Trafficking

13.2.1 Definition and Elements of Human Trafficking

One of the current and most globally accepted definitions of human trafficking was issued by the United Nations Palermo Protocol. In said Protocol, TIP is defined as "…the recruitment, transportation, transfer, harboring or receipt of persons, by means of the threat or use of force or other forms of coercion, of abduction, of fraud, of deception, of the abuse of power or of a position of vulnerability or of the giving or receiving of payments or benefits to achieve the consent of a person having control over another person, for the purpose of exploitation. Exploitation shall include, at a minimum, the exploitation of the prostitution of others or other forms of sexual exploitation, forced labor or services, slavery or practices similar to slavery, servitude or the removal of organs" (UN Palermo Protocol, 2001). Based on the given definition, human trafficking consists of three elements: the act, the means, and the purpose. The act defines what is done (recruitment, transportation, transfers, etc.), the means determines how the act is done (by threat or use of force, coercion, etc.), and the purpose defines why the act is done using a particular means (exploitation including prostitution of others, sexual exploitation, etc.). The United Nations Office on Drugs and Crime (UNODC) makes a clear differentiation between human trafficking and migrant smuggling, terms frequently confounded. The UNODC states four main differences between human trafficking and migrant smuggling. First, migrant smuggling involves consent while a victim is trafficked against his/her consent. Second, migrant smuggling ends with the migrant's arrival at their destination, whereas trafficking involves the ongoing exploitation of the victim. Third, smuggling is always transnational, while trafficking can take place within state's borders. Finally, sources of profits in smuggling cases derive from the facilitation of illegal entry or stay of a person in another country, while in the case of trafficking, profits are derived from exploitation.

13.2.2 The Victims and the Traffickers

According to the UNODC Global Report on TIP 2016, approximately 17,000 victims were detected in 85 countries in 2014. As shown in Figures 13.1 and 13.2, Western and Central Europe is the region where most human trafficking victims are detected followed by North America, Central America, and the Caribbean. Among them, around 70% were female—50% women and 20% girls. Even though the majority of the victims were female, the share

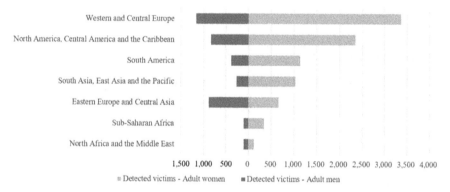

FIGURE 13.1
Detected adult victims. (Source: Global Report on TIP 2016 (UNODC).)

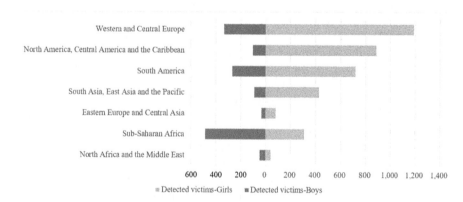

FIGURE 13.2
Detected children victims. (Source: Global Report on TIP 2016 (UNODC).)

of male (men and boys) victims as well as girls has doubled since UNODC started collecting data on TIP in 2003. The share of men by region ranges between 15% and 34% with the exception of Sub-Saharan Africa (9%) and Eastern Europe and Central Asia (53%).

Child victims account for 25%–30% of all cases and vary significantly by region. For instance, 64% of cases detected in Sub-Saharan African countries are children while wealthier countries of North America, Europe, and the Middle East report around 20%–25% of cases. Factors that facilitate human trafficking in children include cultural practices, low accessibility to education, lack of institutions dedicated to child protection, and weak criminal justice systems.

Victims detected depend on the form of exploitation. While the majority of male victims are linked with trafficking for forced labor, female victims are usually trafficked for sexual exploitation. Overall, most (53%) of the victims detected are trafficked for sexual exploitation; in particular in Western and Southern Europe and Central and South-Eastern Europe. On the other hand, trafficking for forced labor is more pronounced in Eastern Europe and Central Asia.

The same report that discusses traffickers (Global report on trafficking in persons, 2016) show the majority (about 60%) of traffickers are men. The trend applies to all regions except Eastern Europe and Central Asia, where the majority of offenders—as well as victims—are women. Participation of women in human trafficking activities is remarkably high compared to other crimes and is increasing. Persons involved in criminal activities related to human trafficking can be differentiated as investigated, prosecuted, or convicted depending on their legal status. Approximately 15,500 persons were investigated for TIP around 2014, 8,500 were prosecuted, and 3,300 were convicted.

TIP can be carried out by members of criminal organizations, by couples—husband/wife or girlfriend/boyfriend—or by family members. Criminal organizations work predominantly at an international level while couples and family members operate domestically. Traffickers have usually been victims themselves of human trafficking. In many cases, especially in sexual exploitation, victims are forced to recruit other trafficking victims.

13.2.3 Risk Factors for Trafficking

Both push and pull factors create the conditions for human trafficking on which traffickers rely. While anecdotal reports suggest many of these factors, few have stood up to rigorous statistical testing. These factors include conflict, statelessness, refugees, religious minorities, poverty, lesbian, gay, bisexual, transgender, queer (LGBTQ) identification, disabilities, runaway youth, and indigenous populations. Policymakers and survivor service providers must consider these largely nebulous factors. Cho used (2015) extreme bound analysis (EBA) to account for push and pull factors of human trafficking in destination and origin countries and found difficulty in identifying robust variables and reliable, empirical data. This study both statistically corroborates and refutes several of the potential underlying causes of human trafficking mentioned above. Nonetheless, this represents the first and only attempt to statistically model and predict factors of human trafficking. Understanding these and other underlying causes to human trafficking provides the foundation upon which more complex cognitive models can be designed and tested.

13.2.4 The Stakeholders' Perspectives

Counter-trafficking stakeholders include local, domestic, and international governments; law enforcement; academia; the technology sector; nonprofit/nongovernmental organizations; and commercial enterprise. Currently, only law enforcement and government see themselves as hunters and responders—that is, organizations capable and mandated to find traffickers and victims and do something to disrupt their activities. Using the approaches highlighted below in Sections 13.3 and 13.4, nonprofit and commercial enterprises may also become empowered to hunt and respond to trafficking in novel ways.

13.3 Data Science and Geographic Information Systems in Human Trafficking

13.3.1 Analytic Platforms and Techniques

Current efforts to utilize technology are limited but have had substantial successes at reducing the time it takes to gather evidence, identify victims, and provide national services. Human trafficking hotlines, such as the ones operated by Polaris in the United States and Stop the Traffik in the United Kingdom, offer substantial opportunities to learn about trends in trafficking as well as the ability to create and distribute useful analytics to share with other organizations. MEMEX[1] was a dispersed project that developed tools to be deployed on the dark web through the Tor[2] network. Recently, some of the more useful tools developed as part of MEMEX have been updated and distributed through academia.

Significant data-based efforts to combat trafficking have developed out of Thorn, Marinus Analytics, and Global Emancipation Network. Nonprofit Thorn's Spotlight tool aggregates the U.S.-based sex trafficking ads

---

[1] www.darpa.mil/program/memex.
[2] www.torproject.org/.

throughout the open web and allows law enforcement agencies to extract ads based on search criteria, view images, and build evidence files. Marinus Analytics' tool, Traffic Jam, offers similar ability on a for-profit basis including some image analysis capability. Global Emancipation Network, a nonprofit, aggregates data from around the globe on all forms of human trafficking and allows stakeholders to cohost relevant data on their analytic platform, Minerva, which allows for identifying correlation among disparate datasets. Other tools on Minerva include multiple image analysis technologies, blockchain and cryptocurrency analysis, public records enrichment, natural language processing, network analysis, and more.

Stakeholders need the ability to engage multiple analytic techniques on a seamless user interface that allows them to interact with data across time and geography. From time-scrubbing visualizations to interactive map analysis, hours and lives can be saved with the introduction of robust visualization and modeling tools.

13.3.2 Geographic Information System Visualization Tools in Human Trafficking

Data mined by anti-trafficking organizations from social networking websites reflects the interactions between actors or the movement of actors that are potentially part of the human trafficking industry. These networks are naturally represented as mathematical graphs, where a graph is defined as a set of nodes and edges between nodes. Interactions or movements are directional connections. Hence, a directed graph, G, can be used to describe the network, where $G = (N, A)$, such that N is a set of nodes and A is a set of arcs, which are ordered pairs of elements of N. In social media data, an arc (or directed edge) can represent an exchange, such as a message sent from one actor (source node) to another (destination node). Further information about the nature of each interaction, such as the quantity or date and time, can be stored with each edge as a set of attributes.

Visualizations are important for exploring the relationships in these datasets, since node and edge count can be large. Many graph drawing algorithms rely on shifting the node positions to optimize aesthetic properties, such as minimizing edge crossings, overall graph area, or bends in edges (Hu and Nöllenburg, 2016; Herman et al., 2000). Graph connections can also be displayed with a matrix if no meaning is attached to node position. However, human trafficking networks are spatially anchored with geographic locations.

Cartographic flow maps represent network data in its spatial context. Flow maps display the movement of people or goods with arrows whose thickness depends on the magnitude of flow. Phan et al. (2005) used Catmull–Rom spline-based edges to generate the flow maps in the style of Minard's 1864 French wine export map. To achieve this effect, node positions are modified, while preserving relative distance or horizontal and vertical relationships to other nodes, nodes are clustered hierarchically, and edges are rerouted based on bounding boxes around point clusters. Verbeek et al. (2011) further refined this style with techniques to bundle edges to reduce edge crossing. The output is elegant, though bundling edges and clustering points does reduce the level of detail available in the visualization. Boyandin et al. (2010) used flow maps to study migration over time. Here, instead of arrows, a red to green diverging color ramp on the edges displays the direction, with red for origin and green for destination. Migration data tends to be unidirectional, naturally reducing bidirectional clutter. Certain smuggling datasets (movements crossing international borders) might generally share this property with migration, but edge coloring may be less effective for trafficking (within country) data.

Geographic information system visualization tools can be leveraged to support human anti-trafficking efforts. Ibanez and Suthers (2014) tracked phone numbers in online escort advertisements over a 6-week period. A phone number being advertised in different cities is likely to indicate trafficking, as traffickers seek fresh markets for a victim. The results, displayed on maps with arrows, show a west coast circuit, an east coast circuit, and a bicoastal circuit. Standard GIS symbology available with desktop GIS tools (e.g., ArcGIS Desktop or QGIS), effectively communicates these results, but custom GIS visualization tools are needed to support exploratory analysis of large datasets.

GIS web-mapping application programming interfaces (APIs) can be used to create tools designed for anti-trafficking investigations. JavaScript API for ArcGIS, OpenLayers, and Leaflet, popular web-mapping platforms, offer similar functionality, while the JavaScript API for ArcGIS is proprietary, the others are both open source. Proprietary GIS tools require license purchase, but the documentation is more extensive than the open-source equivalents. At the same time, open-source GIS platforms do reduce barriers for sharing tools and analytic techniques among partners and open source has broad support in the GIS community.

Figure 13.3 shows an example of a web-mapping application built to support exploring potential human trafficking data extracted from social media sites. HTVIZ[3], in Figure 13.3, built with OpenLayers and other open-source JavaScript libraries (node, mongoDB, plotly, and d3), visualizes a directed graph, G, that represents network data (nodes and edges) on the map. The user can pan and zoom to explore the graph within a geographic context and select edges and nodes to inspect their attribute values.

The data used in this application was scraped from an online escort service by the Global Emancipation Network. The edges represent aggregated customer reviews of services provided by escort service workers. Reviews are likely to correspond to in-person meetings for service provision. Since interdiction often occurs at major highway truck stops, the path between

---

[3] http://go.ncsu.edu/htviz.

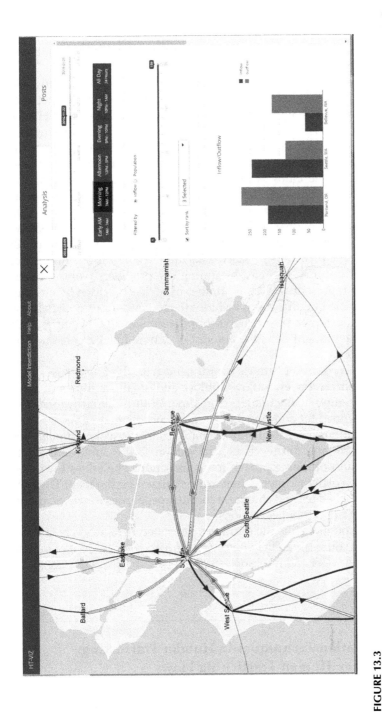

FIGURE 13.3

HTVIZ shows escort message board reviews on a map and in charts on the right. Users can filter that reviews based on time and cities. Here, inflow edges to the cities selected in the drop-down filter are outlined and shown in light gray on the map. Two of the three selected cities are shown here on the map. The dual bar chart shows a city's inflow (the number of reviews written for workers in that city) and outflow (the number of customers writing reviews in that city).

reviewer city A and reviewee city B is not necessarily shown on the map as a direct link. Instead, the network shortest path algorithm is used to compute any intermediate cities and the edges are routed through the cities included on the shortest path. The thickness of each edge depends on the number of reviews routed along that path.

Since node position has real meaning, the graph can't be deformed to avoid edge crossings, but one effective way to reduce clutter in complex network visualizations is to enable the user to interactively filter the data. HTVIZ enables filtering based on review timestamps and cities. The map is updated based on the user selections. Graphs that provide additional views of the data in the panel on the right are also updated based on the user selections.

In Figure 13.3, three cities Seattle, Washington; Portland, Oregon; and Bellevue, Washington were selected. Two of these cities, Seattle and Bellevue, are visible on the map. The inflow edges to these two are outlined and light gray. The inflow–outflow bar chart on the right panel in Figure 13.3 shows two bars for each selected city. The inflow bar signifies the total number of reviews received by providers in a city. This might represent the number of interactions that occur in this city, assuming the providers work at home. The outflow bar represents the total number of reviews written by customers in that city. Note that this count includes the reviews for providers in the same city as the customer, plus reviews for providers in all other cities. This number might be thought of as the amount of demand being generated by residents of this city.

HTVIZ also includes other charts coordinated with the filters: a heatmap matrix to summarize review inflow/outflow and small multiples of sparklines to show temporal fluctuation over time. Built modularly, tools like these can be extended for additional filtering or coordinated views.

Similar applications have been built to enable users to invoke a model (like the ones described later in this chapter in Section 13.4) to analyze the data and determine an optimal approach for interdiction. The edges identified by the model as worthy of interdiction are visualized on the map. Then the user can also select edges and change associated properties to see the model output under different input conditions.

GIS visualization tools like the examples described here, when connected to real-time data sources, can help investigators discover evidence of human trafficking in their data and make decisions about staging interdictions.

13.4 Optimization Techniques in Human Trafficking: Disrupting Human Trafficking Flows

Human trafficking and, in general, illicit networks can be modeled as a graph where *nodes* represent the entities that form the network and *arcs* denote the relationship or connection between them. Example of *nodes* in a human trafficking network include traffickers, intermediaries, and buyers. Human trafficking victims can be considered the "resource" that flows through the network. *Network optimization* solves problems that can be structured as networks. Examples of problems traditionally solved with network optimization include the shortest path, maximum flow, and the network interdiction problems. *Network interdiction* problems belong to a family of more sophisticated network problems. Such models are designed to find a subset of arcs that must be destroyed (i.e., interdicted) in order to, for instance, maximize the shortest path or minimize the maximum flow. In this section, we describe two interdiction problems in human trafficking.

There is clearly a high level of uncertainty in modeling human trafficking networks. Uncertain parameters in human trafficking networks include demand, supply, prices and distribution channels. There are primarily two ways that uncertainty is typically incorporated in a mathematical optimization model. One way, which is called *stochastic optimization*, models uncertainty directly by using probability distributions to describe uncertain parameters. However, solutions to such models assume these distributions to be correct, which might not be true in some cases. Another approach is *robust optimization (RO)*. In *RO*, it is only necessary to describe a range of possible parameters (i.e., the "uncertainty set") without knowing the underlying true probability distribution. Solutions to these types of problems are "robust" in the sense that they are the best solution assuming any possible scenario in the uncertainty set could take place. The clandestine nature of human trafficking networks makes it difficult to gather enough useful and reliable information to develop a probabilistic model. In addition, *stochastic* models tend to require more computational resources than *robust* models. Thus, we believe that *RO* is a suitable approach to model uncertainty in human trafficking networks.

In some applications, decisions do not all need to be taken at the beginning of the decision process (here-and-now) but can be taken once the uncertainty is realized (wait-and-see). For instance, in a classical ISE example, if we are dealing with the problem of how much to produce at a plant and how much of a particular product to distribute to the retailers given uncertain demand, we can easily see two types of decisions. On one hand, the production-type decisions (how much to produce) need to be taken here-and-now due to the required time-consuming production process. On the other hand, the distribution-type decisions (how much to ship to each retailer) can be taken once the demand is realized, i.e., wait-and-see, given a quick distribution time. Based on the temporal nature of the decisions, robust models can be divided into two categories: if all decisions are assumed to be required here-and-now, the robust model is called *static*. If some decisions can be taken later (wait-and-see), the robust model is called *adaptive*.

In this section, we provide two robust network interdiction models where flow—the number of human trafficking victims moved through the network—is considered uncertain. The first model, a *static* robust formulation, determines the links to interdict in the network in order to minimize flow by using a limited number of resources required for interdiction. The second model, an *adaptive* robust formulation, determines the links to interdict in the network and the number of resources required in order to restrict the flow to a desired level at minimum cost.

13.4.1 Network Interdiction

In an interdiction problem, a leader partially or fully destroys arcs of the follower's network in order to block the follower's flows, delay the delivery length of a supply, or decrease the follower's profits (Lim and Cole, 2007). Interdiction models provide ways to infer the vulnerabilities of a particular system (Dimitrov and Morton, 2013). Applications for interdiction problems range from purely military and security-related problems such as drug trafficking and interdicting enemy's supply lines to non-military applications such as preventing the spread of an infectious disease (Lim and Cole, 2007) or interdicting the electric power grid in order to assess its vulnerabilities (Dimitrov and Morton 2013). Cartwright (2000) mentions supply chain interdiction as a possible operational strategy in order to disrupt the competitor's supply chain. Interdiction actions can be interpreted in a wide variety of ways and depend on the scenario in which the interdiction takes place. In a military context, air strikes or checkpoints might be considered as interdiction actions while in a public health scenario, the location of sensors that can detect a threat of a particular kind might be the interdiction action of choice.

13.4.1.1 The Max Flow Interdiction Problem

As found in Wood (1993), the max flow interdiction problem (MFIP) can be described as follows: given a directed $G = (N, A)$ with a source and terminal nodes s and t, respectively, a capacity u_k for each arc $k \in A$, a resource need r_k for interdiction of any arc $k \in A$, and an interdiction budget R, which is the subset of arcs $A' \subset A$ whose interdiction does not exceed R that minimizes s–t flow?

The MFIP can be modeled as follows:

$$\text{Minimize} \sum_{(i,j) \in A} u_{ij} \beta_{ij} \tag{13.1}$$

$$\text{Subject to } \alpha_i - \alpha_j + \beta_{ij} + \gamma_{ij} \geq 0, \ \forall (i,j) \in A \tag{13.2}$$

$$\alpha_t - \alpha_s \geq 1, \tag{13.3}$$

$$\sum_{(i,j) \in A} r_{ij} \gamma_{ij} \leq R, \tag{13.4}$$

$$\alpha_i \in \{0,1\}, \forall i \in N \tag{13.5}$$

$$\beta_{ij}, \gamma_{ij} \in \{0,1\}, \forall (i,j) \in A \tag{13.6}$$

Notation can be found in Table 13.1. In essence, the above model identifies an s–t cut where some arcs are broken (i.e., $\gamma_{ij} = 1$) and the remaining arcs in the cut are used (i.e., $\beta_{ij} = 1$) so as to leave as little available capacity as possible. In particular, Eq. (13.1) minimizes the capacity of the arcs left unbroken, Eqs. (13.2) and (13.3) are flow constraints and Eq. (13.4) represents the budget constraint.

In the model defined by equations (13.1)–(13.6), a single-source single-terminal model, can be adapted to multiple sources and nodes by creating artificial nodes s and t and adding appropriate indestructible arcs. Furthermore, budget can be converted into a cardinality budget where R is the number of arcs that can be interdicted. Additional extensions such as multiple interdiction resources and multiple commodities can be found in Wood (1993).

13.4.2 Robust Optimization

One of the standard methods for modeling uncertainty is RO. RO is a methodology which is concerned with finding solutions that perform well with respect to uncertain future conditions (Peng et al., 2011). For a comprehensive review in RO, see Ben-Tal et al. (2009), Bertsimas et al. (2011), and Gabrel et al. (2014). RO is particularly useful in situations where it is very difficult to identify probability distributions to model the uncertain data (Ben-Tal et al., 2009). The complexity and unique characteristics of illicit network interdiction makes this problem a suitable candidate to exploit the benefits of RO.

Initial robust models such as Soyster (1973) and Bertsimas and Sim (2004), only modeled "here-and-now" decisions without considering that some

TABLE 13.1

Notations Used in MFIP, RMFIP, and ARMFIP Models

Index	
k	$\text{arc}(i,j) \in A$
Decision Variables	
α_i	1 if node i is on the t side of the cut and 0 if node i on the s side of the cut.
α_j	1 if node j is on the t side of the cut and 0 if node i on the s side of the cut.
β_{ij}	1 if arc (i,j) is a forward arc across the cut but it is not broken, 0 otherwise.
γ_{ij}	1 if arc (i,j) is a forward arc across the cut which is to be broken, 0 otherwise.
Parameters	
u_{ij}	Capacity of arc (i,j)
r_{ij}	Cost of interdicting arc (i,j)
R	Interdiction budget

decisions might be made after the uncertainty is realized. Such models are called *static* models. The general *static* model considered is

$$\min_{x \in \mathcal{X}} \max_{u \in \mathcal{U}} f(\mathbf{u}, \mathbf{x}) \tag{13.7}$$

where decision $\mathbf{x}$ must be feasible for all possible realization of uncertain parameter $\mathbf{u}$ (i.e., all realizations of $\mathbf{u}$ that belong to the uncertainty set $\mathcal{U}$).

Recent robust models, such as Ben-Tal et al. (2004) and Dunning (2016), consider the notion of sequential decision-making (i.e., *adaptive*) which is the idea that all decisions do not have to be taken "here-and-now," but some decisions can be made after the realization of the uncertain parameter. For instance, a two-stage problem in which we first make a decision $\mathbf{x}$, then uncertain parameters $\mathbf{u}$ are realized, and then we make another decision $\mathbf{y}$ can be written as

$$\min_{x \in \mathcal{X}} \max_{u \in \mathcal{U}} \min_{y \in \mathcal{Y}} f(\mathbf{u}, \mathbf{x}, \mathbf{y}). \tag{13.8}$$

This problem can also be expressed as

$$\min_{x \in \mathcal{X}, y(u)} \max_{u \in \mathcal{U}} f(\mathbf{u}, \mathbf{x}, \mathbf{y}(\mathbf{u})), \tag{13.9}$$

where second-stage decisions $\mathbf{y}(\mathbf{u})$ are said to be functions of the uncertain parameter $\mathbf{u}$.

Given its intractability, efforts to solve the adaptive RO problem have focused on developing approximation methods. For a detailed discussion on such methods, the interested reader is referred to Dunning (2016).

13.4.3 A Robust Interdiction Model for Disrupting Illicit Human Trafficking Networks

Consider a set of pairs $(i, j) \in A$ that represent the arcs in the graph $G = (N, A)$ as described in the previous section. Let arc capacities u_{ij} be subject to uncertainty. Each u_{ij} is modeled as a symmetric and bounded random variable $\tilde{u}_{ij}$ that takes values in $\left[u_{ij} - \hat{u}_{ij}, \ u_{ij} + \hat{u}_{ij} \right]$, where $\hat{u}_{ij} = \epsilon_{ij} u_{ij}$. u_{ij} is the nominal value of the flow while ϵ_{ij} is the flow variability. Additionally, we limit the number of uncertain parameters that can vary by using the budget uncertainty set proposed by Bertsimas and Sim (2004) (i.e., $\left\| \tilde{u}_{ij} \right\|_1 \leq \Gamma$). The robust MFIP (RMFIP) can be formulated as follows:

$$\text{Minimize } Z \tag{13.10}$$

$$\text{Subject to } \sum_{(i,j) \in A} u_{ij} \beta_{ij} + z\Gamma + \sum_{(i,j) \in A} p_{ij} \leq Z \tag{13.11}$$

$$z + p_{ij} \geq \hat{u}_{ij} y_{ij}, \ \forall (i, j) \in A \qquad (13.12)$$

$$-y_{ij} \leq \beta_{ij} \leq y_{ij}, \ \forall (i, j) \in A \qquad (13.13)$$

$$z \geq 0; \ p_{ij}, \ y_{ij} \geq 0, \ \forall (i, j) \in A \qquad (13.14)$$

Constraints (13.2–13.6)

where z, p_{ij}, and y_{ij} are auxiliary variables, and Γ is the maximum number of uncertain parameters that can vary. One of the main advantages of the budget uncertainty set model proposed by Bertsimas and Sim (2004) is that the level of conservativeness can be controlled by changing the value of the parameter Γ. If $\Gamma = 0$, none of the uncertain parameters are allowed to change, thus, the solution with the nominal values will be found (i.e., no protection). On the other hand, if $\Gamma = |N|$, where N is the set of uncertain parameters, all the uncertain parameters are allowed to change: in this case, the solution found is the most conservative (i.e., Soyster's model).

13.4.4 A Resource-Allocation Adaptive Robust Model for Disrupting Human Trafficking Flows

In addition to determining which arcs to interdict (i.e., γ_{ij}), in the following adaptive robust interdiction model we want to know the number of resources (e.g., law enforcement agents, checkpoints) required to be assigned to each interdicted arc under arc capacity uncertainty.

Consider a set of pairs $(i, j) \in A$ that represent the arcs in the directed graph $G = (N, A)$ as described before. The objective is to reduce the total interdiction cost given that flow must be restricted to a predetermined value F. Each resource used in the interdiction plan reduces the capacity of the arc where it is located by h units. Since arc capacities are uncertain, allocated resources might be higher than the "true" arc capacity. Therefore, idle resource capacity is penalized by a parameter π. Resources at interdicted arcs can be located "here-and-now" (i.e., uncertainty is still present) at a cost c^1 or can be added once the arc capacity is realized (i.e., wait-and-see) at a cost c^2. Resources located "here-and-now" are represented by the variable x_{ij}^1 while resources located later are represented by the variable $x_{ij}^2(\mathbf{u})$ and are functions of the uncertain parameter $\mathbf{u}$ (i.e., arc capacity).

The adaptive RMFIP (ARMFIP) can be formulated as follows:

$$\text{Minimize } z \qquad (13.15)$$

$$\text{Subject to } \sum_{(i,j) \in A} c^1 x_{ij}^1 + c^2 x_{ij}^2(\mathbf{u}) + I_{ij}(\mathbf{u}) \geq z, \ \forall \mathbf{u} \in \mathcal{U} \qquad (13.16)$$

$$I_{ij}(\mathbf{u}) \geq 0, \ \forall (i, j) \in A, \ \forall \mathbf{u} \in \mathcal{U} \qquad (13.17)$$

$$\sum_{(i,j)\in A} u_{ij}\beta_{ij} \leq F, \tag{13.18}$$

Constraints (13.2–13.3)

$$x^1_{ij},\, x^2_{ij}(\mathbf{u}) \in Z^+,\, \forall (i,j) \in A,\, \forall \mathbf{u} \in \mathcal{U} \tag{13.19}$$

Constraints (1.5–1.6)
  where

$$I_{ij}(\mathbf{u}) = h * \left[x^1_{ij} + x^2_{ij}(\mathbf{u}) \right] - u_{ij}\gamma_{ij} \tag{13.20}$$

The uncertain arc capacity u_{ij} is modeled as before with a level of conservativeness $\Gamma = |N|$.

When there are multiple sources and multiple sinks, constraints must be added such that $\alpha_i = 0$ for all sources and $\alpha_i = 1$ for all sinks.

One approach to solve adaptive robust models is called *linear decision rules* (LDRs) or the *affine adjustable robust counterpart* (AARC) proposed by Ben-Tal et al. (2004). Despite its popularity, it is limited to continuous second-stage variables. Since we are assuming that resources are integers, we use the adaptive mixed-integer optimization approach (AMIO) proposed by Dunning (2016) which allows to solve robust problems with integer second-stage variables.

Although benefits of using adaptive compared to static policies are problem specific, preliminary experiments using the adaptive policy found with model (ARMFIP) achieved up to 37% improvement in the objective value compared to the corresponding static policy and up to 44% when comparing both policies under simulated scenarios.

13.5 Conclusions

Human trafficking can be thought of as modern-day slavery; and while this seems unfathomable, at least 20 million people are believed to be enslaved around the world (International Labour Force, 2017). While there have been numerous efforts to thwart offenders at all levels, ranging from local to national and international, the focus has primarily been on policy and public health measures. Identifying and prosecuting offenders is difficult because, among other reasons, they operate in clandestine networks. In this chapter, we introduced two concepts that could enhance efforts to protect victims and prosecute offenders; visualization techniques to help identify victims and discover trafficking networks as well as mathematical modeling using

RO to allocate resources to disrupt these networks. Both methods deal with uncertainty of information and complement each other.

For future efforts to succeed, the academic community needs to form strong partnerships with NGOs and law enforcement organizations that are forging revolutionary data collection and extraction efforts. With these partnerships, visualization tools and optimization models can be harnessed to make discoveries within these big data troves. Geovisualizations and analytics fed by these resources can be used to search for geospatial patterns that could indicate the movement of individuals or groups, potential victims, and any temporal rhythms to these patterns. Graph visualization techniques such as edge bundling and relaxing node placement constraints can be used to display more complex graphs. The design of new dashboards with maps and information visualizations should be driven by the questions analysts are asking.

RO is a promising and relatively new way of dealing with optimization-under-uncertainty. Exploring new algorithms and methods for solving RO problems, in particular AMIO is still necessary. Approaches for solving AMIO problems, such as the partition-and-approach (Dunning, 2016) and the "Adjustable Robust Mixed-Integer Optimization Method via Iterative Splitting of the Uncertainty Set" (Postek and den Hertog, 2016) can serve as starting points to develop more efficient algorithms.

Inevitably, a gap is created when theoretical models are proposed to solve real-world problems such as human trafficking. The development of user-friendly decision support systems that combine visualization tools and optimization models can effectively close the gap between theory and practice. Building on concepts like those described in this chapter, the ISE/OR community is poised to make valuable contributions to the fight against human trafficking.

Acknowledgments

This work supported was in part with funding from the Laboratory for Analytic Sciences (LAS). Any opinions, findings, conclusions, or recommendations expressed in this material are those of the authors and do not necessarily reflect the views of the LAS or any agency or entity of the U.S. government.

References

Baer, Kathryn. Debate-the trafficking protocol and the anti-trafficking framework: Insufficient to address exploitation. *Anti-Trafficking Review* 4 (2015).

Ben-Tal, Aharon, Alexander Goryashko, Elana Guslitzer, and Arkadi Nemirovski. Adjustable robust solutions of uncertain linear programs. *Mathematical Programming* 99, no. 2 (2004): 351–376.

Ben-Tal, Aharon, Laurent El Ghaoui, and Arkadi Nemirovski. *Robust Optimization*. Princeton, NJ: Princeton University Press, 2009.

Bertsimas, Dimitris, David B. Brown, and Constantine Caramanis. Theory and applications of robust optimization. *SIAM Review* 53, no. 3 (2011): 464–501.

Bertsimas, Dimitris, and Melvyn Sim. The price of robustness. *Operations Research* 52, no. 1 (2004): 35–53.

Boyandin, Ilya, Enrico Bertini, and Denis Lalanne. Using flow maps to explore migrations over time. In *Geospatial Visual Analytics Workshop in conjunction with the* 13th *AGILE International Conference on Geographic Information Science*, Guimarães, Portugal, Vol. 2, no. 3. 2010.

Buchin, Kevin, Bettina Speckmann, and Kevin Verbeek. Flow map layout via spiral trees. *IEEE Transactions on Visualization and Computer Graphics* 17, no. 12 (2011): 2536–2544.

Cartwright, Shawn D. Supply chain interdiction and corporate warfare. *Journal of Business Strategy* 21, no. 2 (2000): 30–35.

Chisolm-Straker, Makini, and Hanni Stoklosa, eds. *Human Trafficking is a Public Health Issue: A Paradigm Expansion in the United States*: Springer, 2017.

Cho, Seo-Young. Evaluating policies against human trafficking worldwide: An overview and review of the 3P index. *Journal of Human Trafficking* 1, no. 1 (2015): 86–99.

Dimitrov, Nedialko B., and David P. Morton. Interdiction models and applications. In *Handbook of Operations Research for Homeland Security*, pp. 73–103. New York: Springer, 2013.

Dunning, Iain Robert. Advances in robust and adaptive optimization: Algorithms, software, and insights. PhD dissertation, Massachusetts Institute of Technology, 2016.

Gabrel, Virginie, Cécile Murat, and Aurélie Thiele. Recent advances in robust optimization: An overview. *European Journal of Operational Research* 235, no. 3 (2014): 471–483.

Gozdziak, Elzbieta M., and Elizabeth A. Collett. Research on human trafficking in North America: A review of literature. *International Migration* 43, no. 1–2 (2005): 99–128.

Herman, Ivan, Guy Melançon, and M. Scott Marshall. Graph visualization and navigation in information visualization: A survey. *IEEE Transactions on Visualization and Computer Graphics* 6, no. 1 (2000): 24–43.

Hu, Yifan, and Martin Nöllenburg, eds. *Graph Drawing and Network Visualization: 24th International Symposium, GD 2016*, Athens, Greece, September 19–21, 2016, Revised Selected Papers. Vol. 9801. Springer, 2016.

Ibanez, Michelle, and Daniel D. Suthers. Detection of domestic human trafficking indicators and movement trends using content available on open internet sources. In *2014 47th Hawaii International Conference on System Sciences (HICSS)*, pp. 1556–1565. IEEE, 2014.

International Labour Organization (ILO). www.ilo.org/global/lang--en/index.htm, 2017.

Konrad, Renata A., Trapp, Andrew C., Palmbach, Timothy M., and Blom, Jeffrey S. Overcoming human trafficking via operations research and analytics: Opportunities for methods, models, and applications. *European Journal of Operational Research* 259, no. 2 (2017): 733–745. doi:10.1016/j.ejor.2016.10.049.

Lim, Churlzu, and J. Cole Smith, Algorithms for discrete and continuous multicommodity flow network interdiction problems. *IIE Transactions* 39, no. 1 (2007): 15–26.

Peng, Peng, Lawrence V. Snyder, Andrew Lim, and Zuli Liu. Reliable logistics networks design with facility disruptions. *Transportation Research Part B: Methodological* 45, no. 8 (2011): 1190–1211.

Phan, Doantam, Ling Xiao, Ron Yeh, and Pat Hanrahan. Flow map layout. In *IEEE Symposium on Information Visualization, 2005. INFOVIS 2005*, pp. 219–224. IEEE, 2005.

Postek, Krzysztof, and Dick den Hertog. Multistage adjustable robust mixed-integer optimization via iterative splitting of the uncertainty set. *INFORMS Journal on Computing* 28, no. 3 (2016): 553–574.

Soyster, Allen L. Convex programming with set-inclusive constraints and applications to inexact linear programming. *Operations Research* 21, no. 5 (1973): 1154–1157.

UN Palermo Protocol. United Nations Protocol to prevent, suppress and punish trafficking in persons, especially women and children, supplementing the United Nations Convention against transnational organized crime, 2001.

United Nations Office on Drugs and Crime (UNODC). Global report on trafficking in persons. www.unodc.org/unodc/data-and-analysis/glotip.html, 2016.

Wood, R. Kevin. Deterministic network interdiction. *Mathematical and Computer Modelling* 17, no. 2 (1993): 1–18.

14

The Role of Manufacturing Process Design in Technology Commercialization

Brian K. Paul and Patrick McNeff
Oregon State University

Sam Brannon
Hewlett-Packard Inc.

Michael O'Halloran
CH2M

CONTENTS

14.1 Introduction

Over the past 20 years, manufacturing has reemerged as a frontier for industrial and manufacturing engineering research within the United States (US). This reemergence has been spurred on, in part, by expanding federal budgets in manufacturing research extending from an awareness of the importance of manufacturing to our economy (PCAST 2012). The rising of manufacturing research has accompanied a concurrent desire to advance university knowledge into the marketplace aided in large part by entrepreneurs. Experience in working with entrepreneurs to commercialize technology at Oregon State University (OSU) has shown that entrepreneurs generally lack the skills needed to develop manufacturing strategies for taking new products to market. In this chapter, we use the term "Manufacturing Process Design" (MPD) to describe the unique engineering activity which manufacturing engineers perform to support product development, in contrast with other definitions of MPD involving the optimization of existing manufacturing processes (Rhyder 1997). Further, we differentiate MPD from manufacturing systems design involving the development of planning and control strategies for governing the manufacturing process (Black 1991).

Below, we provide some background on microchannel process technology (MPT), which is used as a context for describing MPD in the sections that follow.

14.2 Microchannel Process Technology

MPT are energy systems and chemical processing equipment that relies on embedded submillimeter microchannels for their function. In order for the microchannel processing of mass and energy to be industrially viable, large arrays of microchannels (Figure 14.1c) are needed to scale up volumetric flow rates as follows:

$$\dot{v} = V_{\text{avg}} \cdot A \tag{14.1}$$

where $\dot{v}$ is the volumetric flow rate of reactants through microchannels, V_{avg} is the average velocity of the reactants through the microchannels, and A is the flow cross section which is a product of the flow cross section of each microchannel and the number of microchannels.

MPT is considered process intensification technology in that microchannels reduce the footprint of heat exchangers, reactors, and separators (Tonkovich et al. 1998, Brooks et al. 1999, Hong 1999, Warren et al. 1999). This intensification is due to (i) the shorter diffusional distances within microchannels leading to accelerated heat and mass transfer performance and less surface area per device, and (ii) the higher surface-area-to-volume ratio of microchannel arrays (Ameel et al. 1997, 2000).

Early examples of MPT in applications of energy systems and chemical systems included microelectronic cooling systems (Little 1990, Kawano et al. 1998), chemical reactors (Martin et al. 1999a,b, Matson et al. 1999), fuel processing systems (Tonkovich et al. 1998, Daymo et al. 2000), and heat pumps (Drost and Friedrich 1997, Drost 1999), among many others. Figure 14.1a shows a microchannel solar receiver (MSR) that is one-fifth the weight of alternative solar receivers, allowing for smaller, less costly central towers as a part of a heliostat power plant.

Industrial acceptance of MPT has been slow largely due to the high cost of MPT components. The size of microchannel features requires the use of microchannel lamination or microlamination strategies (Paul and Peterson 1999, Paul 2006). Microlamination consists of the patterning and bonding of thin layers of material, called laminae, to produce monolithic components with embedded microchannel features. Challenges lie in the breadth

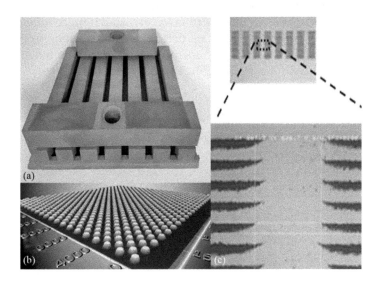

FIGURE 14.1
(a) Oblique view of a 15 cm MSR; (b) scanning white light interferometric image showing an obliquie view of a micropin array that is embedded within the down facing surface of the MSR; (c) a cross section of a microchannel array produced by diffusion bonding. (Courtesy of Patrick Scot McNeff.)

of dimensional integration required with the most undesirable effect being variation in the thermal or reaction conditions across the channel array caused by the maldistribution of flow due to poor manufacturing tolerances.

14.3 Manufacturing Process Design

MPD is the activity of generating, evaluating, and specifying the flow of process steps needed to produce a particular product at a required cost. The definition of a process step is shown in Figure 14.2. Process steps involve a workpiece, a piece of capital equipment to carry out the process function on the workpiece, and touch tools, such as cutting tools or workholding tools, that directly *touch* the workpiece and through which the process function is carried out. The machine tool is designed to implement the process parameters needed to impart shape or material properties to the workpiece. Touch tools are used to directly impart a new shape to the workpiece or to align the workpiece within the machine tool coordinate system. A final MPD involves the specification of the flow of process steps (process flow) needed to produce a product and the specification of the details of each process step including the machine tool, process parameters, touch tooling design, and workpiece specification. Details of process parameter optimization and touch tool design are described in other textbooks on material processing and tool design. The focus of this chapter is on determining the proper process flow for a product within a market and identifying the machine tools, feasible process parameters, and cost elements necessary to implement and evaluate that process flow. In this chapter, we equate this activity to MPD.

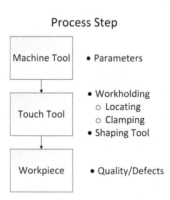

FIGURE 14.2
Definition of a process step.

14.3.1 Process Requirements

Like all design activities, MPD starts by establishing a set of requirements. *Manufacturing process requirements* are the technical and business requirements that must be satisfied by a MPD. Three sets of requirements are considered for MPD. First, the product being developed has performance requirements and a set of conditions that expand into material and geometric specifications by the product designer. These requirements include dimensional and geometric tolerances which are typically communicated by the product designer in the form of engineering models and drawings. Second, the product must be produced at a cost target for a particular market size that enables the entrepreneur to make a profit. It is recognized that different markets require different cost targets and have different market sizes. The manufacturing engineer will want to start with a product that allows for the highest cost target and the largest market demand. Market demand determines the volume of production which is a key factor in driving down the cost of components. Third, downstream processes often set requirements for upstream processes. Examples include brazing clearances or joint configurations for welding that must be machined or otherwise formed in an upstream step. Together, these process requirements are used to drive process specifications in the form of a manufacturing process flow diagram (PFD).

14.3.1.1 Engineering Specifications

The engineering specification provides the first and most obvious set of manufacturing process requirements. The product designer converts the functional requirements of the product into a set of *material and geometric specifications* communicated as engineering drawings and models. Engineering drawings are typically organized into a set of assembly, subassembly, and component drawings showing the mechanical relationship of each component to one another. Further, the engineering specification will include an engineering bill of materials which describes the parts, part quantities, and material specifications necessary to build and assemble the component. The material chosen will often be the cheapest that meets the design requirements for mechanical rigidity and any product-specific material properties, such as thermal properties or oxidation resistance.

In addition to material specifications, dimensional and geometric tolerances for each step must be determined. Process tolerances fall into two categories: in-process and finished. In-process tolerances are tolerances required by a future step in the process, such as a parallelism tolerance in sheet stock to be diffusion bonded to ensure even pressure distribution or a brazing allowance between two parts to be brazed to enable capillarity between parts. Finished tolerances should be the same as those of the final product and are typically specified at the component level.

For example, most microchannel devices consist of arrays of micro-channels where channel variations of greater than 5% relative standard deviation can lead to flow maldistribution between channels and poor effectiveness (Paul 2006).

14.3.1.2 Production Cost Targets

Production costs are the costs associated with all required elements for manufacturing a product. This includes costs for capital equipment and capital facilities as well as labor, consumables, utilities, and equipment maintenance. Using accounting vernacular, the production cost is an early estimate for the cost of goods sold (COGS). For a product to be economically viable, production cost estimates must be well below the price of the product in the market. This is because the price of the product must include the costs of distribution, marketing, profit, and administrative overhead. For the average mechanical product, a rule of thumb is that the price is typically about three times the COGS and the COGS is about three times the raw material.

At the beginning of a design effort, cost targets are typically fluid in that there are not hard targets at which people will no longer purchase a product. Business techniques can be used to assess what the distribution of prices would be for a product in a particular market. However, more typically, MPDs and cost targets are reiterated simultaneously until the business team feels a case can be made to enter the market. During these reiterations, the cost breakout by percentage across cost categories and process steps is used to determine areas to focus on in redesign efforts to reduce costs.

Example 1: Process Requirements for the MSR

The MSR is a solar thermal microchannel panel which sunlight is focused on the surface and a supercritical CO_2 working fluid takes transfers the heat to a turbine for energy generation. The max operating conditions for the MSR is 760°C at 250 bar internal pressure. To meet the MSR's 95% thermal efficiency requirements, the 300 µm-deep microchannels can vary by no more than 5%. In addition, a pressure drop of 10 bar must not be exceeded in the device
Process requirements are as follows:

- Material must have sufficient strength at 760°C at 250 bar internal pressure.
- Wall thickness must be sufficiently thick to withstand 250 bar at 760°C given material choice.
- Final channel depth can only vary by less than 15 µm.
- The material must be corrosive resistant.
- Distribution and microchannel flow paths must not exceed 10 bar pressure drop.

14.3.2 Conceptual Design

Prior to detailed design, initial efforts are needed to propose a *manufacturing process concept,* which is described as the initial process flow required to build the component. Conceptual design involves the selection of process steps for all component-level parts and for all assembly, subassembly, and joining operations necessary to produce the final part. The concept is captured in a manufacturing PFD.

14.3.2.1 Manufacturing PFD

The *manufacturing PFD* is a flow diagram showing the order of process steps required from raw material to final product. A PFD for an MSR Figure 14.3 is shown in Figure 14.4. Process steps are represented by boxes and connected by arrows to show the flow of the workpiece. Single flow shows steps in each part to be manufactured. A separate, orthogonal flow shows the sequence of assembly or joining steps needed to produce the final product. In Figure 14.4, four vertical flows represent the four parts of the MSR, while the single horizontal flow shows how the parts are joined together. Note that each process step is numbered to simplify downstream analyses. Circular nodes between blocks are used to document process requirements. These nodes are useful in showing where critical product tolerances are affected at different stages of the build. Process requirements also prescribe the need for inspections or statistical methods to ensure product performance and downstream manufacturability.

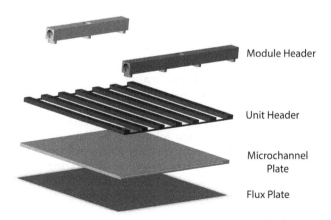

FIGURE 14.3
Alpha prototype of the MSR including machined distribution system.

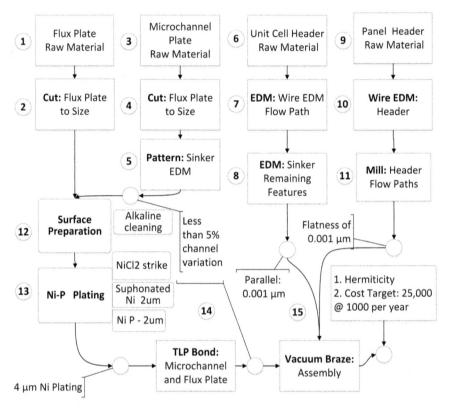

FIGURE 14.4
Initial PFD for the MSR.

Example 2: PFD of the MSR

Based on a preliminary interaction with vendors, an initial PFD was developed for the MSR. At the top of the diagram, the raw material for each component part flows into a sequence of process steps that moves down and to the right. The final device is on the bottom right. To complete the diagram, it must be both hermetic and economically feasible. If either of these conditions is not met, another iteration of the PFD should be performed, attempting to either reduce cost or increase production reliability. In-process tolerances are called out by orange circles that represent pass–fail inspection stages. An example is after step 11 a flatness of 0.001 μm must be met on the brazing surface for a successful braze.

14.3.2.2 Manufacturing Process Selection

Within any given manufacturing process step, there may be several ways that step can be implemented. For example, steps 2 and 4 in Figure 14.4 are both blanking steps for sheet metal in which the raw material sheet is cut to the size of the lamina that is needed. Blanking can be implemented using photochemical machining (PCM) or with a punch and die. Which process you choose should reflect the process requirements. For example, during early stages of product development, the device will need to be prototyped to confirm that the product concept meets functional requirements. The criteria used to select process steps for prototyping, with a production volume of one, are different than the criteria used to select process steps for manufacturing thousands or millions. Criteria for prototyping include the speed with which the job can be turned around as well as the cost of the touch tooling for making the part. The cost and lead time to make the punch and die set are prohibitive during prototyping. Consequently, PCM would be a better choice for blanking during prototyping while the punch and die would be a better choice assuming that the process requirements include a sizeable market to offset the costs of the die set during production.

However, other process requirements must also be considered beyond just cost as a function of production volume. As described above, the process requirements in a manufacturing setting also require meeting the dimensional and material requirements of the part. In the example above, a well-controlled PCM process could likely provide more delicate features than the punch-and-die process. Further, if the processes are being implemented in an expensive material, such as a Ni-based superalloy, forming may be preferred over machining to improve material utilization. When producing a small number of parts in a very cheap material, machining may be preferred over forming. Below are some general criteria to consider in the initial stages of conceptual design.

Component tolerances: It is important to remember that capabilities of any process chosen must be capable of satisfying the dimensional and geometric tolerances of the part.

Expensive material: The criteria in process selection are different if the MPD is driven by material costs versus processing costs. For products driven by material costs, net shape or near net shape processes like casting, molding, and forming processes, which are capable of higher material utilization, should be considered.

Production volume: At low production volumes, the cost of specialized machine tools and work holding tools cannot be distributed over enough products to be cost effective. Consider moving from net shape and near net shape processing toward additive and subtractive processes with more general-purpose tooling.

Geometric features: Additive manufacturing has the capability to produce parts with complicated internal structures that would normally require several parts to be joined. Additive manufacturing has actually been found to be superior in market sizes beyond tens of thousands when (i) it enables features that provide significant performance advantages and (ii) it reduces the number of components and joining operations required.

Example 3: Mechanical Machining versus Green Sand Casting of the MSR

The initial prototype distribution system for the MSR described above was fabricated using machining because standard diameter pipes that meet pressure vessel standards were not readily available for Haynes 230. This machining of the headers resulted in very low material utilization. To improve this material utilization, green sand casting was investigated as described in more detail in the case study at the end of the chapter.

Normally green sand casting would not be considered for microchannel applications, but the header system was found to be sufficiently large enough to allow for its use. Adjustments to the design of the headers were made to implement the header as a single part casting. Use of green sand casting required the design of two green sand cores to implement the hollow internal features of the header system. Support of the cores was not a concern as the base of the flow path is open allowing for the cores to be placed into the green sand mold parting line prior to placing the cope on the drag. Walls of the headers were sufficiently thick that vendor expected that the melt will have good flow through the mold cavity. To meet tolerances after casting, a grinding step was added, so that the header assembly could provide brazing clearances of 0.001 in.. The new MSR panel concept showing the cast header is shown in Figure 14.8. The updated PFD is shown in Figure 14.9. Table 14.1 shows the comparison of total cost of raw materials for each of the designs.

14.3.3 Detailed Design

Detailed MPD involves selecting specific machine tools to carry out the requirements for each process step. Machine tool specification and selection are important factors in determining the overall cost of a product. Capital equipment costs are driven by the capacity and capability of the machine tools selected. For example, higher tonnage presses cost more than lower tonnage presses, and a 30 cm work envelope typically costs more than a 15 cm work envelope. Further, capital facility costs are driven by the machine tool

TABLE 14.1
Material Cost for Alpha Design and Updated Design

Design	Distribution Raw Material Cost ($)	Material Cost Reduction (%)
Alpha prototype	23,000	N/A
Manufacturable design	12,395	48

footprint. Utility costs are driven by machine tool power requirements. The number of laborers is directly related to the number of machine tools specified. Consumables can be driven by machine tool selection such as certain furnaces which may require inert gases. Machine tool selection to a large degree locks in the processing costs of the product, ultimately affecting the shape of production cost curves and determining the price of the product in the market. The specification and selection of machine tools for each process step is determined based on the capability and capacity requirements of each process step.

14.3.3.1 Capability Analysis

A *capability* is the ability of a machine tool to carry out a particular processing function. For example, presses must administer force in forming process steps, furnaces must control temperature profiles, and machining centers must provide certain cutting power during machining process steps. In many cases, the capability of a machine tool is governed by multiple criteria. For example, stamping operations may require presses with not only a maximum load but also adequate displacement and perhaps control over strain rate. Heat treatment operations may require furnaces that control the ramp rate in both heating and cooling as well as the type of environment. Finally, the capability of the machine tool depends on the material and processing interaction and can usually be predetermined based on the required outcome. For example, stamping of a geometry into a sheet metal part in stainless steel will require higher forces than in aluminum.

The capability of a machine tool is different from the ability of a workpiece or touch tool to survive the operation. For example, the elongation of the material being deformed governs deformation processes. Machining operations are constrained by the temperature that the cutting tool can withstand. So to reduce the risk of specifying a manufacturing solution that cannot be attained, machine tool specification may be accompanied by manufacturability analyses to determine that the machine tool, touch tool, and workpiece combination being proposed is viable. Once machine tools have been specified, efforts are needed to work with vendors to identify and select machine tools capable of meeting that specification at the lowest cost.

Many manufacturing materials and processing textbooks exist that can help with specifying the capability of machine tools (Groover 2007, Kalpakjian and Schmid 2014, Black and Kohser 2017). These textbooks also can help with the accompanying manufacturability analyses of the workpiece and touch tools in order to determine the viability of a production step.

14.3.3.2 Grouping Process Steps for Tool Sharing

Once the machine tool capabilities for each process step have been determined, efforts can be made to group processes requiring similar capabilities in an effort to reduce the capital cost of equipment. If several process steps are capable of sharing the same machine tool, then tool sharing will reduce the number of machine tools, requiring less capital. For example, assume that three laminae must be produced to implement a particular reactor design. One of the laminae has a load requirement that is an order of magnitude less than the other two. Several options are possible. At low production volumes, all three can be placed onto the same press reducing the number of presses that must be purchased. Alternatively, at higher production volumes, it might be more economical to purchase two presses, with one processing the two laminae at higher loads and the other processing the lamina at lower loads. The outcome of this activity is a *preliminary machine tool specification*. The next step is to determine if what the capacity of these machine tools need to be in order to satisfy market demand.

Example 4: Grouping Stamping Operations

A steam reforming system has been designed to produce hydrogen for a fuel cell by reacting steam and a hydrocarbon fuel at high temperatures. It is desired to take the residual heat in the reformate stream and move it into the reactant stream prior to reaction to make the reaction more efficient. A waste heat recuperator is designed to do so, consisting of two sets of shims interdigitated for a total stack of 53 shims per heat exchanger. The forces required for embossing the two laminae were calculated to be 67 and 257 kN. To reduce the capital cost at low production volumes, the two process steps can be combined allowing one press to be purchased capable of providing a maximum force above 254 kN.

14.3.3.3 Capacity Analysis

Capacity (or cycle time) analysis is a critical part of specifying machine tools. The *capacity* of a machine tool is the number of parts that can be produced in 1 year. Capacity can be calculated by dividing the total number of hours available for that tool in a year by the cycle time (in hours) required for one part. The capacity is an integer number so it is rounded up to the next whole number. The cycle time should include the time needed to physically process one workpiece on the machine tool plus the time spent loading and unloading the part onto the machine tool.

Example 5: Calculating the Capacity for a Vacuum Hot Press

One way to make a metal microchannel array is to stack together patterned metal laminae and diffusion bond them into a monolith using a vacuum hot press (VHP). The VHP is a hydraulic ram with platens that can apply pressure inside a hot zone existing within a vacuum vessel. The stack of laminae is placed on platens within the hot zone and heated to the bonding temperature. At the bonding temperature, metal atoms in the adjoining laminae surfaces within the stack move by solid-state diffusion across surface boundaries producing metallic bonds. A typical diffusion bonding cycle requires a short time to load the laminae within a bonding fixture followed by a heating ramp to the bonding temperature at which time the hydraulic pressure is applied. After several hours of solid-state diffusion at the bonding temperature, the bonding pressure is released, and the sample is cooled yielding a monolithic microchannel array. For this calculation, assume the following:

Loading time = 0.1 h
Heating ramp = (1,025°C–25°C)/(10°C/min) = 100 min = 1.67 h
Bonding time = 4 h
Cooling ramp = (1,025°C–125°C)/(5°C/min) = 180 min = 3 h
Unloading time = 0.15 h

$$\text{Cycle time} = \sum_1^N t_n = 8.92 \text{ h}$$

where t_n is the time needed for each substep in the process (e.g., loading, ramping). To determine the capacity of the VHP, the total number of hours that the VHP can operate must be calculated as follows:

Total hours available in a year

$$= 24 \text{ h/day} \times 7 \text{ days/week} \times 52 \text{ weeks} = 8{,}736 \text{ h}$$

Total hours available per year by the vaccum hot press

$$= 8{,}736 \times 80\% \left(\text{due to scrap and downtime}\right) = 6{,}988.8 \text{ h}$$

where the 80% is calculated as the total uptime (82.5%) times the total quality yield (97%). Consequently, the capacity of the VHP can be calculated as follows:

$$\text{Capacity} = \frac{\text{VHP hours available}}{\text{Cycle time per part}} = \frac{6{,}988.8}{8.92} = 783.5 \text{ stacks} = 783 \text{ stacks}$$

Interpreting these results, the VHP is capable of producing 783 microchannel arrays per year. The implication of this analysis is determined by comparing the capacity of the machine tool with the required production volume. In this case, if the required production volume was 1,000 arrays per year, two VHPs would be required to set up the factory.

The goal of cycle time analysis is to try and design each machine tool setup so that it slightly exceeds the required production volume. This is because, as previously noted, the number of machine tools drives all other processing cost categories. The estimated COGS for a process step cannot go lower than the point at which the machine tool makes the maximum number of parts. If all process steps are targeted to be at maximum utilization at the required production volume, then the COGS for the product cannot go lower than this cost, even if the production volume increases. This is because an increase in the production volume will trigger the procurement of more machine tools for each step which will actually cause the COGS to increase. Continued increases in production will actually cause a decrease in the estimated COGS, but the COGS will never go below the COGS at maximum (80% in this case) equipment utilization.

In reality, it is not possible to get all machine tools to maximum utilization at the required production volume. This is because there are not an infinite number of machine tool capacities for each process step. Each family of machine tools typically has a discrete number of capacities. As suggested above, one strategy to consume excess capacity in a machine tool involves combining the production requirements of several process steps with similar capability requirements. Consider Table 14.2 showing the stamping operations needed for laminae required to build a recuperative heat exchanger. Note that at the original production volumes of interest, the utilization of the machine tool in each process step is at most 4.1% suggesting excessive capacity. Purchasing one press capable of meeting all force requirements will result in a reduction of capital costs, increasing the percent utilization of the production press to 28.5%.

More typically, efforts are needed to increase the capacity of a machine tool. Attributes of machine tools that can lead to an increase in capacity include, but are not limited to, the work envelope and the production rate. For example, if the diffusion-bonded array above is made of laminae that are 50 mm × 50 mm, procurement of a VHP with a 100 mm × 100 mm work envelope would increase the capacity of the tool by four. Further, tripling the spindle power of a machining center would increase the material removal rate by three times. The implication of capacity considerations is that proper sizing of machine tools dictates where the knee in the production cost curve is, which is a key factor in MPD for minimizing cost at required production volumes.

To more fully understand how the work envelop of a machine tool can be used to size the capacity of a machine tool, consider two different types of process steps: serial versus parallel processes. *Serial processes* are those that are executed sequentially, as in laser processing, while *parallel processes* are those processes that are executed simultaneously, such as diffusion bonding (Rhyder 1997). The use of larger work envelopes within serial processes has much less effect than on parallel processes. Consider the following examples.

Within a given process flow containing all parallel processes, it may not be possible to find the same sized work envelope for all machines. This leads to considerations of depaneling and singulating. Consider the processing of a printed circuit heat exchanger (PCHE) using PCM and diffusion bonding

TABLE 14.2

Capability and Capacity Analyses for Stamping Process Steps at 100,000 units/year

Substep Number	Unit	Process Name	Capability Force (N)	Required Cycle Time (s)	Tool Capacity (parts/year)	Target Number of Unit	Required Production Volume	Total Tool Hours Required (Tool-h)	Available Tool (h/year)	Tool Count at 1 unit/cycle	Tool Share (%)
2.1	Reformate shim		24,969.00	0.2	18,000		2,700,000	—	4,608		
2.2			229,035.00	0.2	18,000		2,700,000	—	4,608		
5.1	Reactant ship	Blank emboss	254,004.00	0.2	18,000	3,049,838	2,700,000	169.4	4,608	3.68%	17.48
5.2		Blank emboss	67,667.00	0.2	18,000	2,936,881	2,600,000	163.2	4,608	3.54%	16.84
12.1	Top bottom plate	Top bottom plate blanking	165,970.00	3	1,200	222,525	200,000	185.4	4,608	4.02%	19.14
12.2		Top plate interconnect hole punching	12,746.00	3	1,200	109,568	100,000	91.3	4,608	1.98%	9.42
13.1	Reformate header	Header blanking	107,445.00	3	1,200.	219,137	200,000	182.6	4,608	3.96%	18.84
13.2		Header bending	546.5	3	1,200	212,562	200,000	177.1	4,608	3.84%	18.28
		40 metric ton press tool	392,000.00							21.03%	100.00

Substep numbers beyond the decimal indicate a progressive tool setup on the machine tool. Rightmost gray areas show percentage utilization of machine tool and percentage that each step will use the tool. Bottom gray area shows the machine tool capability needed to satisfy the steps specifying a 40 metric ton press.

(Johnston 1983). The work envelope for PCM may be as large as 48 in. wide while the work envelope for the next step in the process flow, diffusion bonding, is constrained by the size of the VHP which generally can accommodate panels that are 24 × 24 in.. For sake of illustration, assume that a particular PCHE is just under 24 × 24 in. in size. When the initial mother panel is machined, four shims (laminae) are etched in parallel onto one 48 × 48 in. panel. However, the resulting panels are found to be too big for the work envelope of the downstream VHP. To accommodate the VHP, the panels are depaneled, or blanked, into four sets of 24 × 24 in. laminae that can be diffusion bonded in the VHP. The final PCHE would need to be singulated from the 24 in. bonded panel stack after diffusion bonding. In this case, depaneling of a single layer panel could be carried out by laser machining, waterjet machining, or sheet metal blanking while singulating may involve a much thicker bonded panel stack which would require sawing or wire electrodischarge machining (EDM).

Increasing the work envelope of parallel processes provides a direct multiplier for increasing the capacity of the process. In example 5, the 48 in. PCM line has double the capacity of the 24 in. line. Further, a 24 in. VHP would have four times the capacity of a 12 in. VHP. In other words, running the same diffusion bonding cycle in the 24 in. VHP produces four times the number of 6 in. bonded panel stacks than the 12 in. VHP could produce. In contrast, increasing the work envelope of serial processes does not have a multiplying effect. Consider replacement of the PCM step with a laser cutting process. Given the same laser power output, the process is constrained by the scan speed of the laser. A larger work envelope can provide small incremental increases in capacity by reducing the amount of loading and unloading time per part.

Example 6: Metal Additive Manufacturing

To highlight the importance of maximizing the use of the work envelope within a particular process step, consider the additive manufacturing of metals. For laser powder bed fusion tools, there are two primary operations: powder layering and laser cladding. Powder layering involves adding a thin layer of metal powder onto the powder bed. Laser cladding involves scanning the laser over the powder bed to melt and densify the metal, allowing it to adhere to the surface beneath. Both process steps are serial and are repeated over and over again to produce the final part. Consider two build approaches. The first build cycle is used to make one part whereas in the second build cycle is optimized to arrange the build of four parts simultaneously. The layering time needed in both cases is identical, and in the second case, that layering time is spread over four parts. However, the laser cladding involves scanning the laser over the entire cross section of each part. Consequently, the cladding time takes up a larger proportion of the cycle time for the second scenario (Table 14.3).

TABLE 14.3

Cycle Time Comparison One versus Four 3D Printed Parts per Cycle Time

Description	One Part Build Cycle (h)	Four Part Build Cycle (h)
Cladding time	3.1	12.4
Layering time	8.9	8.9
Total time per part	12	5.3

The outcome of detailed design is a *final machine tool specification* that can be used to produce a layout for manufacturing operations. This typically requires interactions with vendors to learn what capabilities and capacities exist for performing certain process steps. This type of vendor interaction requires the specification of both the capability and capacity requirements of a machine tool before vendor engagement as shown in Table 14.3. Sometimes, the required machine tool may not be available, which will require adjustment in the capability or capacity of the process step to perform the necessary operation within the constraints of the machine tools available. An example of a final machine tool specification is shown below for the application of sheet metal stamping. In this case, the machine tool was sized by force requirements and the excess capacity was chosen because the client was interested in production costs across a range of production volumes.

Inevitably, as the MPD unfolds, it may be found that production costs exceed the cost target. In such a case, it is important to realize that some process steps will be more important than others, so efforts must be made to evaluate which process steps have the most influence on COGS. In this manner, it may not be important to size the capacity of all machine tools to a given market. Rather, the emphasis should be on adjusting the capacity of the steps having the largest influence on cost.

14.3.3.4 Machine Tool Flow Diagram

The machine tool flow diagram is an outcome of the machine tool specification above. The diagram shows how material flows into and out of different machine tools represented as aggregates of process steps with similar capabilities. This allows for a process designer to determine whether there may be material flow challenges associated with the machine tool arrangement between shared tools in upstream and downstream processes. To create a machine tool flow diagram, all process steps with similar capabilities from the manufacturing PFD are grouped together into a single machine tool. For machines that are near maximum capacity, setup time should be considered to ensure adequate capacity.

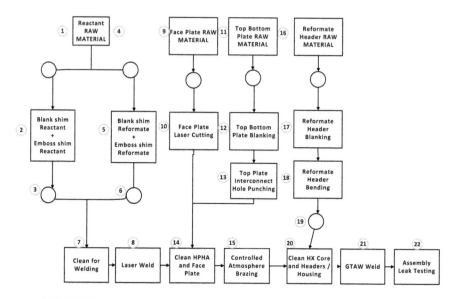

FIGURE 14.5
Manufacturing PFD for a laminated heat exchanger.

Example 7: Machine Tool Flow Diagram

Figures 14.5 and 14.6 show an example of a manufacturing PFD and its corresponding machine tool flow diagram, where like capabilities are grouped in order to determine the set of machine tools that need to be modeled in design evaluation.

14.3.4 Design Evaluation

In early product development, evaluation of the MPD will be mainly with respect to cost targets. In order to evaluate the MPD relative to cost targets, a *production cost model* is needed, capable of estimating the COGS for the product. In this case, the cost model is built from the bottom up by adding the raw material costs to the processing costs for each process step. Processing costs are determined based upon a process step analysis involving six cost elements, several of which are dependent on production volume.

The model is developed assuming that the factory will be established as a greenfield site, i.e., no existing building or capital equipment. This assumption is a good starting point for entrepreneurs with no existing assets. At lower production volumes, the COGS obtained with the cost model is typically more expensive than what can be produced in the supply chain. This is because suppliers typically have higher labor and equipment utilization due to the ability to supply multiple customers. However, at higher production volumes, the investment of capital can be depreciated over a larger number of products making the COGS estimation cheaper than what can be obtained

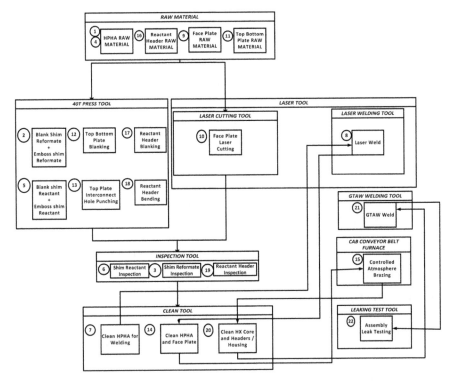

FIGURE 14.6
Machine tool flow diagram showing the grouping of process steps with the same capability.

in the supply chain. However, this assumes that the tacit process knowledge needed to operate the process under control exists within the organization building the greenfield plant. The cost model is assembled by adding together the raw material and processing cost elements (labor, equipment, maintenance, facilities, consumables, and utilities) for each process step. A detailed description for how to calculate each processing cost element can be found elsewhere (Gao et al. 2016).

14.3.5 Cumulative Yield

Cumulative yield is an important concept for cost modeling. Each step has some yield less than 100% and, consequently, is required to produce more components than scheduled to meet the annual demand. For example, consider the bonding of two laminae with a yield of 90%. If 10 bonded assemblies are needed, then 11 assemblies must be bonded to offset the yield. In addition, this concept compounds over many steps so that upstream processes (those at the start of the process flow) have to make up for losses in downstream processes. This loss of yield must translate into additional costs within the cost model and is captured by a cumulative yield from the current process step, x, to the end of the process flow, N.

$$Y = \prod_{i=x}^{N} y_i, \tag{14.2}$$

where Y is the cumulative yield and y_i is the yield of every downstream step in the PFD including the current process step.

To aid in the calculation of the production cost, a MATLAB program was developed to make all raw material and processing calculations based on the format of an input spreadsheet. The format of the input data is shown in Table 14.4. The resultant calculations are plotted in a production cost curve as shown in Figure 14.11. The characteristic curve is an exponential decay with asymptotic behavior that moves toward minima as the utilization of all machine tools in the MPD is maximized. The production volume at which the curve approaches the minimum production cost is considered the knee in the curve. In Figure 14.11, the curve appears to reach near minima between 1,000 and 2,000 units/year.

If the production cost at the production volume of interest is not below the cost target, additional design work is needed. Development of a strategy for taking the next iteration of MPD typically involves producing pie charts showing whether material or processing costs are dominating. In the latter, efforts are needed to determine which process steps are most important and, then, which cost elements are driving those process steps. In our experience commercializing MPT, several strategies have been used for reducing costs. First, if material cost dominates, efforts must be made to move toward near net shape, high material utilization processes like casting and forming. Second, if processing costs dominate, efforts must be made to determine if wholly new manufacturing technologies can be used to address cost drivers. Some of these lessons are expressed in the following case study.

14.4 Case Study: Microchannel Solar Receiver

The U.S. Department of Energy is interested in finding ways to replace coal-fired power plants with solar-driven power plants. Typical power plants operate on a Rankine cycle where a boiler superheats steam which is used to drive a turbine that produces electricity. The output of the turbine is mixed gas and liquid. In order to pump this fluid back to the boiler, additional heat must be removed through the use of condenser heat exchangers. Once the fluid is restored to liquid phase, it can then be pumped back up to the boiler, restarting the cycle.

TABLE 14.4

Process Input Database Example for Calculating Production Costs

Category	Parameters	Units	Description	Top Bottom Raw Material	Top Bottom Plate Blanking	Top Plate Interconnect Hole Punching
Raw material	M_y,p	kg/device	Mass of part material per device	0.15447357	0	0
	M_m,p	$/kg SS 439	Cost of material per unit*3% scrap rate	5.6753	0	0
	P_share	%	% share tool utilization	100.00%	19.14%	9.42%
Tool	T	$/tool	Capital tooling cost		$325,000.00	$325,000.00
	T_i	% of capital	Tool installation cost		10%	10%
	y_t	years/tool	Amortized life of tool		10	10
	n_dy	Devices/year	Annual demand		200,000	100,000
	Y_t	%	Yield of manufacturing process		80%	80%
	U_t	%	Maximum utilization of tool		80%	80%
	h_y	h/year			4,608	4,608
	C_t	Shim/h-tool	Tool capacity = 1/cycle time		1,200	1,200
	n_DY		*n_DY*	100,000	100,000	100,000
Facility	N_t	# of tools	Number of tools		1.00	1.00
	B_A	$/m²	Building cost per m²		1,000	1,000
	K_bt		Tool footprint multiplier		3	3
	a_t	m²	Tool footprint*2 for storage		14	14
	y_b	years/building	Amortized life of building		30	30
	n_dy	Devices/year	Annual demand		200,000	100,000

(Continued)

TABLE 14.4 (*Continued*)

Process Input Database Example for Calculating Production Costs

Category	Parameters	Units	Description	Top Bottom Raw Material	Top Bottom Plate Blanking	Top Plate Interconnect Hole Punching
Labor	N_t	# of tools	Number of tools		1.00	1.00
	L	$/person-year	Annual labor cost		$50,000	$50,000
	R		Average labor loading rate		1.5	1.5
	p_t	People/tool	Number of laborers per tool		1	1
	n_dy	Devices/year	Annual demand		200,000	100,000
Maintenance	N_t	# of tools	Number of tools		1.00	1.00
	e_t	Percent/year	Annual maintenance as % of tool cost		5%	5%
Consumables	T	$/tool	Capital tooling cost		$325,000.00	$325,000.00
	n_dy	Devices/year	Annual demand		200,000	100,000
	c_i	$/Blanking Die	Blanking Die cost		$120,000.00	$120,000.00
	q_c,I	1/(shims/die) = dies/shim	1/blanking die life		3,000,000	3,000,000
	c_m	$/Blanking Die	Blanking Die maint cost		$750.00	$750.00
	q_c,m	1/(shims/die) = dies/shim	1/shim die maint life		400,000	400,000
Utilities	$u\_c,1$	$/kWh *#part	Electricity cost per kWh		0.102	0.051
	$q\_u,1$	kWh/device	Electricity use per device		0.000833333	0.000833333
	$u\_c,2$	$/gal	DI Water cost per gal		0.004	0.004
	$q\_u,2$	gal/device	DI Water use per device		0	0
	$u\_c,3$	$/gal	Wastewater and Sewer cost per gal		0.009	0.009
	$q\_u,3$	gal/device	Electricity use per device		0	0

The idea of a solar-driven power plant is to heat the steam by replacing the boiler with a sun-facing solar receiver designed to transfer heat from the solar flux into the steam. Prior receivers have consisted of an array of centimeter-scale tubes through which the water passes for heating. An MSR is a receiver that has embedded microchannels within the sun-facing surface of the flux plate. Microchannels transfer much more heat from the flux plate to the working fluid than cm-scale tubes. Consequently, the surface area of the MSR can be greatly reduced allowing the solar receiver to be designed with as much as a fivefold reduction in mass. The lighter weight receiver allows the central tower, where the receiver is placed at the focal point of a mirror array, to be reduced in mass and cost.

The MSR would consist of 250, 1 m × 1 m panels placed edge to edge in a circular pattern capable of receiving solar energy from the heliostat array surrounding the tower. The function of a single 1 m × 1 m MSR panel is to distribute the flow of the incoming working fluid down into the microchannel array where it is solar heated and gathered back into a single outflow of superheated fluid. The operating conditions needed for an MSR panel to meet efficiency requirements, using supercritical carbon dioxide as the working fluid, are 760°C and 250 bar internal pressure. The material needed to resist corrosion at these temperatures for 30 years is an expensive Ni-based superalloy, costing five to ten times the cost of steel.

14.4.1 Product and Process Requirements

The product requirements for the MSR included the following:

- Wall thickness must be sufficiently thick to withstand 250 bar at 760°C given material choice.
- The material must be corrosive resistant.
- Distribution and microchannel flow paths must not exceed X bar pressure drop.
- Material must have sufficient strength to withstand 250 bar internal pressure at 760°C.

The initial process requirements for the MSR included the following:

- As-processed material must have sufficient strength to withstand 250 bar internal pressure at 760°C.
- Final microchannel depth must vary by less than 15 μm over an average of 300 μm.

14.4.2 Initial Design

The initial design of the MSR and a manufacturing PFD were shown in Figures 14.3 and 14.4, respectively. This MPD was initially established based on the need to prototype the MSR. Note that in addition to the initial process requirements, the vacuum brazing process required a brazing allowance of 50 μm driving upstream flatness tolerances of 25 μm between the unit cell and panel headers.

14.4.3 Initial Design Evaluation

Using this MPD, efforts were made to develop a cost estimate for the MSR using data from suppliers and equipment vendors. The final cost to produce the MSR prototype (at a production volume of one) was found to be $45,000 per MSR, which did not meet the original cost target of $25,000. To devise a new MPD, the cost of this design was broken down by process step as shown in Figure 14.7 and a strategy for redesigning the MPD was conceived as described in Section 14.4.4.

14.4.4 Detailed Design

Figure 14.7 shows that the dominate cost element is the raw material. The largest mass within the design was found to be within the two headers. Therefore, a key strategy for developing the MPD was to move away from machining processes, which are known to waste raw material via chip removal and move toward a higher material utilization process. Efforts were made to investigate the use of a green sand casting process. Green sand casting is a higher material utilization process with relatively low touch tooling costs in the form of a casting pattern. From a capability standpoint, it was determined that the limiting factor in castability was the ability of the cast melt to fill the mold prior to freezing (avoid a short shot). Discussions

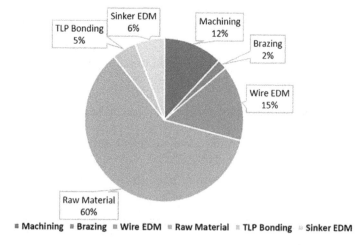

FIGURE 14.7
Cost breakdown of initial vendor estimate by process step.

with vendors determined that for Ni-based superalloys, mold passages, and cavities greater than 25 mm in width would enable mold fill (Figure 14.8). Moving the design of the header system from machining to casting reduced material waste by 48%, a significant cost savings in raw material. Further, because both headers could be cast at the same time, the move toward green sand casting also reduced the number of process steps from 15 to 11, eliminating several vacuum brazing steps (Figure 14.9).

A second consideration in modifying the MPD was relative to the sinker EDM of the post array shown in Figure 14.1b. The issue was that to produce a sinker EDM tool at a scale of 1 m² would require more than six million holes to be drilled into the sinker EDM tool. Reliability of the MSR was strongly linked to having good bonds between all posts and the flux plate, meaning that the yield on holes to be drilled into the sinker EDM tool had to be 100%, which was not deemed feasible. Consequently, the MPD was modified to use a wire EDM approach in which only 6,000 wire line cuts were required to yield a similar post array pattern. It was determined that if one of the wire line cuts was out of tolerance, this was not a case for catastrophic failure.

As a result of the move toward wire EDM, changes were needed in the approach to the transient liquid-phase (TLP) bonding step. The TLP bonding step is performed by electroplating a 4 μm thick NiP layer (4 wt% P) onto the plates to be bonded and heating the plates to 1,150°C. At this temperature, the NiP layers become liquid and the two surfaces are pressed together with 12.7 MPa of bonding pressure for 4 h. The process is considered TLP because as the phosphorous is driven into the material by solid-state diffusion over time, the liquid phase eventually solidifies due to reductions in phosphorous composition causing increases in the material liquidus.

Use of the wire EDM process to produce a post array on the microchannel plate causes the faying surface for the TLP bonding step to be on two levels, i.e., at the top of the posts in the middle of the array and at the bottom of the channel around the perimeter of the array. To compensate for this offset in faying surface height, the flux plate was redesigned with a middle pocket,

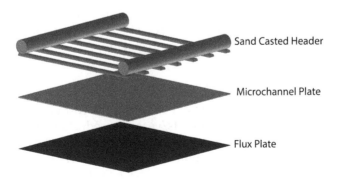

Sand Casted Header

Microchannel Plate

Flux Plate

FIGURE 14.8
Alternative MSR concept with green sand casting distribution system.

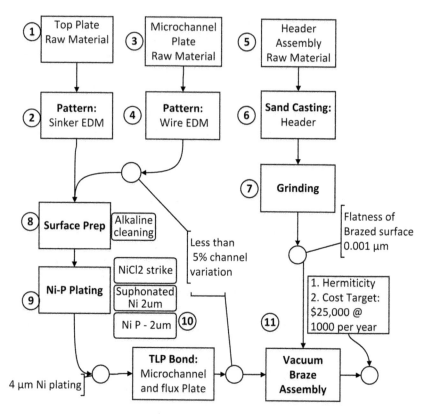

FIGURE 14.9
Updated PFD for the MSR using a green sand cast distribution system.

having a depth to match the offset. When bonding, this allowed two plates to make contact both at the top of the posts in the middle of the array as well as the bottom of the channel around the perimeter of the array. This approach was confirmed by trial with vendors. The final step of the MPD was to vacuum braze the cast header system onto the TLP-bonded microchannel plates using a NiBSi braze foil having a 50 μm thickness.

14.4.5 Final Design Evaluation

Final design evaluation consisted of evaluating the cost of the device at an annual production rate of 1,000 units/year. A production cost model was developed using vendor data along with the methods described in this chapter. This included the cost of an induction furnace for green sand casting of $4 million spread over 10 years. The process adjustments described above allowed for a reduction in the COGS from $45,000 to $23,109 at a production volume of 1,000 units/year which is below the production cost target of

$25,000. A breakout of the final costs is shown in Figure 14.10. The production cost curve for the new MPD is shown in Figure 14.11. This figure shows that the minimum unit cost for the MSR panel asymptotes at around \$21,625 beyond a production volume of about 2,500 units/year. Beyond this production volume, more capital equipment is needed which prevents additional cost reductions based on mass production economics.

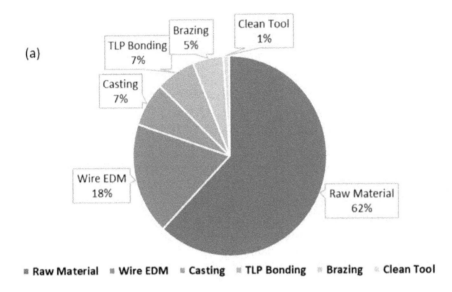

(a)

Raw Material — Wire EDM — Casting — TLP Bonding — Brazing — Clean Tool

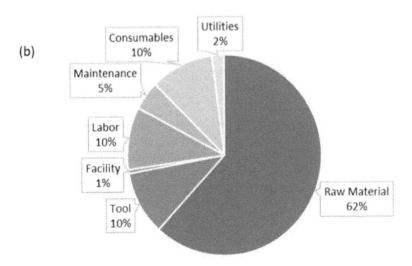

(b)

Raw Material — Tool — Facility — Labor — Maintenance — Consumables — Utilities

FIGURE 14.10
A breakout of the production cost of the MSR by (a) process step and (b) cost category.

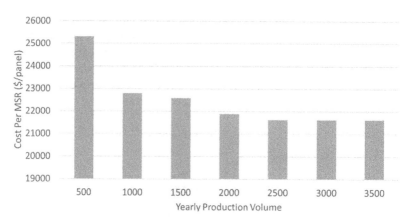

FIGURE 14.11
Cost as a function of yearly production volume.

14.5 Summary

In this chapter, we have demonstrated a framework for MPD used in commercializing product technology at OSU. MPD involves determining process requirement, selecting appropriate processes and machine tools, and evaluating the design based on process-based cost modeling. Several examples are provided from research projects demonstrating how to conduct MPD. A final case study offers an example of how MPD has been used to advance the design of an MSR.

Acknowledgments

This work was supported in part by the U.S. Department of Energy's Energy Efficiency and Renewable Energy SunShot Program under grant number DE-EE0007108. Further, the authors acknowledge the contributions of Kijoon Lee and Chuankai Song, both graduate students of Professor Paul, who collaborated with us to hone and realize the ideas expressed in this manuscript.

References

Ameel, T. A., I. Papautsky, R. O. Warrington, R. S. Wegeng and M. K. Drost (2000). Miniaturization technologies for advanced energy conversion and transfer systems. *Journal of Propulsion and Power* 16(4): 577–582.

Ameel, T. A., R. O. Warrington, R. S. Wegeng and M. K. Drost (1997). Miniaturization technologies applied to energy systems. *Energy Conversion and Management* 38(10): 969–982.

Black, J. T. (1991). *The Design of the Factory with a Future.* New York: McGraw-Hill.

Black, J. T. and R. A. Kohser (2017). *DeGarmo's Materials and Processes in Manufacturing.* Hoboken, NJ: John Wiley & Sons.

Brooks, K. P., P. M. Martin, M. K. Drost and C. J. Call (1999). Mesoscale combustor/evaporator development. *ASME IMECE Conference,* Nashville, TN.

Daymo, E. A., D. VanderWiel, S. Fitzgerald, Y. Wang, R. Rozmiarek, M. LaMont and A. Tonkovich (2000). *Microchannel Fuel Processing for Man Portable Power.* Washington, DC: American Institute of Chemical Engineers.

Drost, K. (1999). Mesoscopic heat-actuated heat pump development. *Proceedings of the ASME Advanced Energy Systems Division Publications AES* 39: 9–14.

Drost, M. K. and M. Friedrich (1997). Miniature heat pumps for portable and distributed space conditioning applications. *Energy Conversion Engineering Conference, 1997. IECEC-97, Proceedings of the 32nd Intersociety,* Honolulu, HI, IEEE.

Gao, Q., J. Lizarazo-Adarme, B. K. Paul and K. R. Haapala (2016). An economic and environmental assessment model for microchannel device manufacturing: Part 1—Methodology. *Journal of Cleaner Production* 120: 135–145.

Groover, M. P. (2007). *Fundamentals of Modern Manufacturing: Materials Processes, and Systems.* Hoboken, NJ: John Wiley & Sons.

Hong, S. (1999). *Experiments and Modeling of a Multiscale Laminar Plate Heat Exchanger.* New York: AES.

Johnston, A. (1983). Printed circuit heat exchangers. Chemeca 83: Chemical engineering today; Coping with Uncertainty. *The Eleventh Australian Chemical Engineering Conference,* Institution of Chemical Engineers, Brisbane, AU.

Kalpakjian, S. and S. R. Schmid (2014). *Manufacturing Engineering and Technology.* Upper Saddle River, NJ: Pearson.

Kawano, K., K. Minakami, H. Iwasaki and M. Ishizuka (1998). Micro channel heat exchanger for cooling electrical equipment. *American Society of Mechanical Engineers, Heat Transfer Division, (Publication) HTD* 361: 173–180.

Little, W. (1990). Microminiature refrigerators for Joule-Thomson cooling of electronic chips and devices. Advances in cryogenic engineering. Vol. 35B-*Proceedings of the 1989 Cryogenic Engineering Conference,* Los Angeles, CA.

Martin, P. M., D. W. Matson and W. D. Bennett (1999a). Microfabrication methods for microchannel reactors and separations systems. *Chemical Engineering Communications* 173(1): 245–254.

Martin, P. M., D. W. Matson, W. D. Bennett, D. C. Stewart and Y. Lin (1999b). Laser micromachined and laminated microfluidic components for miniaturized thermal, chemical and biological systems. *Proceedings of SPIE,* Bellingham, WA.

Matson, D. W., P. M. Martin, D. C. Stewart, A. L. Y. Tonkovich, M. White, J. L. Zilka and G. L. Roberts (1999). Fabrication of microchannel chemical reactors using a metal lamination process. In *Microreaction Technology: Industrial Prospects.* Berlin, DE: Springer: pp. 62–71.

Paul, B. K. (2006). Micro energy and chemical systems (MECS) and multiscale fabrication. In *Micromanufacturing and Nanotechnology.* Berlin, DE: Springer: pp. 299–355.

Paul, B. K. and R. B. Peterson (1999). Microlamination for microtechnology-based energy, chemical, and biological systems. ASME International Mechanical Engineering Congress and Exposition.

PCAST (2012). President's Council of Advisors on Science and Technology, Report to the president on capturing domestic competitive advantage in advanced manufacturing, July 2012, Executive Office of the President.

Rhyder, R. F. (1997). *Manufacturing Process Design and Optimization*. New York: Marcel Dekker.

Tonkovich, A., C. Call and J. Zilka (1998). The catalytic partial oxidation of methane in a microchannel chemical reactor. *AIChE 1998 Spring National Meeting*, New Orleans, LA.

Warren, W. L., L. H. Dubois, S. Wax, M. Gardos and L. Fehrenbacher (1999). Mesoscale machines and electronics--There's plenty of room in the middle. In *Proceedings of the ASME Advanced Energy Systems Division*, Salvador M. Aceves, Srinivas Garimella (Eds.). New York: The American Society of Mechanical Engineers.

15

Improving Responsiveness in Manufacturing Centers through the Virtual-to-Reality Big Data Methodology

Dheeraj Kayarat

Micron Technology Inc.

Todd Easton

Kansas State University

CONTENTS

15.1 Introduction

Companies have benefitted from the massive advancements in computers and technology. Industries generate and store enormous amounts of data. A few new problems now face corporations. "How useful is the gathered big data?" and "How can big data improve the operations of the company?" This chapter discusses these questions and introduces the virtual to reality big data (VRBD) methodology. Virtual reality is frequently abbreviated as VR, and this new methodology requires four Vs and four Rs.

An overview of the VRBD method begins in the virtual world of big data. While no formal big data definition exists, Laney (2001) defines big data in terms of three Vs: volume, velocity, and variability. The goal of most big data projects is to achieve a fourth V, value (LaVelle et al., 2011). The three Rs—remove, recognize, and remedy—analyze the big data to identify and improve productivity. The fourth R, repeat, enables the other three Rs to be rapidly implemented multiple times a day. Thus, the V handles the massive amounts of incoming data in the virtual world, while the R generates value for reality.

VRBD is shown to be effective through an example at a manufacturing facility. Over the past decade, this manufacturing facility spent millions of dollars investing in computational support. Most of this expense can be partitioned into computers, servers, software, sensors, new machines, radio-frequency identification (RFID), and other location identification equipment.

It should be noted that the company did not intentionally invest to achieve big data. Instead, new equipment typically had enhanced monitoring and data reporting. Thus, new equipment also involved the purchase of some computational resources. These resources were fragmented and barely able to communicate. Therefore, the company decided to provide a substantial investment in computational storage with the idea that nearly all data is housed in the same computer cluster.

The facility can now be categorized as a "smart factory" or meeting the trends of Industry 4.0 (Smit et al., 2016). Marr (2016) describes a smart factory as "cyber-physical systems monitor the physical processes of the factory and make decentralized decisions. The physical systems become Internet of Things, communicating and cooperating both with each other and with humans in real time." Automation is at a high level in this facility with material movement handled by automated guided vehicles driven by heuristics. Activities in the factory are managed by sophisticated automated scheduling and dispatching system. Due to the sizeable investment, management's focus is to find benefits from the data to justify the expense. Thus, management's goal is to consistently improve daily operations through big data.

The primary contribution of this chapter is the introduction of the VRBD methodology. Some of the other contributions of this chapter are a discussion of how this method can be implemented to provide benefit to a manufacturing facility. The authors now believe that companies dealing with big data should focus on implementing VRBD to see real-world benefits from the big data.

The remainder of the chapter is organized as follows: Section 15.2 deals with big data and the V portion of VRBD. Section 15.3 introduces the four Rs and explains how they are implemented. Section 15.4 discusses the impact of the VRBD model. Section 15.4 provides the conclusion and topics for future research.

15.2 The Four Vs of Big Data

The importance of big data has permeated society and is used in such diverse areas as healthcare (Raghupathi and Raghupathi, 2014), logistics (Bowersox et al., 2002), and finance (McAfee et al., 2012). The idea of "big" data is nebulous and can range from terabytes to petabytes depending upon the computers and the application of interest. As previously mentioned, Laney (2001) defines big data in relationship to three Vs: volume, velocity, and variability. There is no set formal definition for any of these Vs, but each one creates problems.

Not all big data problems contain all three Vs; however, the application discussed here has all three. This manufacturing facility produces electronic components; has over 5,000 employees and 1,000 machines; and produces about 10,000 parts each day.

Big data's most widely researched V is volume. Volume refers to the amount of data stored and retrieved, which is now ranging in the zetabytes (Sagiroglu and Sinanc, 2013). From Turing's read/write tape (c.f. Garey and Johnson, 2002) to modern cloud computing concepts (Armbrust et al., 2010), storing and retrieving data (Ji et al., 2012; Wu et al., 2013) is a primary research area for the computer science field.

In most instances, a single machine cannot house the data. Much research is performed to determine how to partition the data set so that it is retrievable (Wu et al., 2013). Another vital decision involves the number of copies of data that should be retained to ensure accurate and timely retrieval of the information (Cohen and Shenker, 2002). This redundancy is also important because if a single machine is down, then any request for the data may go unfilled. Thus, multiple copies of the same data are frequently kept.

The manufacturing facility generates approximately 500 GB each day. In 2 days, the facility has generated about 1 TB of data. This data is emanating from over 400 different automated sources. Storing and categorizing this amount of data requires advanced computer science concepts, which are discussed in greater detail later.

Big data's second V, velocity, focuses on the speed at which data is generated and stored. Velocity rarely exists without volume. Faster methods to store or retrieve improve the handling of big data's velocity component. A substantial amount of research on rapidly storing and retrieving data is available, including Assunção et al. (2015), Chinta (2016), and Zaharia (2016).

All of the day's data at the facility is not delivered at midnight. Instead, the data is reported throughout the day. The facility's staging area generates 40,000 rows with 460 data points every minute. This is approximately 150 MB/min. Thus, this facility has velocity.

The third V, variety, involves gathering data from different sources. Even within a single factory, numerous vendors typically provide machines, equipment, and sensors. Thus, the nature of the data is rarely consistent, and yet, the data is important. For instance, one data may be a time stamp on the movement of a forklift, the next data may point to errors identified by a machine, and another data may be the supervisor's report. Clearly, there is little similarity between these types of data. Furthermore, the data may come from different operating systems, such as Windows, Unix, or Solaris. An unsurprising fact is that this facility, like almost every large company, has an enormous amount of variety in its data.

The most common method to deal with variety is through standardization, which is generally a complicated process. Without standardization, data cannot necessarily be combined from different sources, and it can be more difficult to generate improvements. The prevalence and importance of data standardization can be found in Lee et al. (2015) and Garcia et al. (2014). Besides standardization, other researchers focus on how to deal with "messy" data (Milliken and Johnson, 2001). Messy data refers to data that may be categorical or be missing certain portions.

In summary, the example facility has volume, velocity, and variety. Thus, the facility has big data. The goal of most big data projects is a fourth V, value. How can one achieve value in the real world from big data? Fundamentally, value has little to do with the definition of big data. Instead, value focuses on the usefulness of the data. Section 15.3 introduces four Rs that create value for big data projects.

15.3 The Four Rs for Big Data

Value is rarely created by identifying and solving a problem that has already been resolved. Once the problem is solved, some value can be achieved through prevention, but the majority of the benefit has already been achieved. In most manufacturing centers, major errors or events are easily identified by humans. In other words, workers typically see a part that is seriously mangled and the part is discarded. If a second part has a similar characteristic, the process is stopped while the issue is fixed. Similarly, a broken machine or conveyor belt malfunction is almost always instantly identified at a factory. These items are repaired as soon as possible. However, the instruments recording data for these fixed problems do not typically understand these issues and report this data. Since the problems are already resolved, big data should not identify major obvious errors at manufacturing facilities. Rather, big data should focus on more subtle problems. The four Rs—remove, recognize, remedy, and repeat—from VRBD identify less obvious errors.

The first R, remove, deals with the veracity and applicability of the big data. Since machines frequently sense and report data and machines/sensors have error rates, some data may be incorrect. This data should be removed.

Additionally, data related to problems that have already been fixed should also be removed. The goal of remove is to identify and eliminate such data.

Recognize, the second R, is the ability to identify problems and/or opportunities from big data. Additionally, recognize may also report no identified problems and that the status quo is ok. The function of recognize is dependent upon the application and can have various goals. Frequently, recognize focuses on the identification of problem areas. For instance, recognize may identify that a certain machine is producing an above average number of defective parts.

Remedy uses big data to produce corrective measures to the results from the recognize step. In the above example, the recognize step identifies the machine. Remedy examines data to identify trends and suggest root causes of the problem. For instance, examining the historical data for the machine may indicate that the machine has only performed poorly for the past 2 months and during the graveyard shift. The manager of the graveyard shift is notified and meets with a newly hired employee and offers additional training.

These three Rs generate value and benefit from big data. This benefit should not be a one-time occurrence, and the fourth R is repeat. Repeat is the ability to rapidly implement the remove, recognize, and remedy steps. The idea is to move big data projects from off-line to online. Off line problems typically have fixed data, and one seeks a permanent optimal policy. Online problems constantly change data and typically have partially fixed solutions. Thus, no static optimal policy exists. Successfully implementing repeat enables problems to be rapidly identified and solved, which may result in a few changes each day.

In order to repeat, one must focus on algorithms or heuristics that are fast and have a low computational complexity. The three Vs also force all algorithms to have a fast running speed. This is typically achieved by developing heuristics. Heuristics, unlike optimality algorithms, typically evaluate many options and select the best. Heuristics do not guarantee optimality. In contrast, optimality algorithms can obtain a verifiably optimal solution. Even though many optimality algorithms run in polynomial time, many of these algorithms are still too slow for big data.

Since the first three Rs are typically achieved by a heuristical method, researchers constantly develop new and improved heuristics. In this realm, some heuristics are better due to computational speed rather than quality of the proposed solution (Easton and Singireddy, 2008). Thus, anyone implementing big data must balance both the quality and effort required to find the solution.

Although this work is the first to introduce the four Rs in one model, these concepts can be found in other big data research. For instance, Fan et al. (2014) describe the need to remove inconsistent or poorly read data. The concepts of recognize and repeat are easily seen in Veeramachaneni et al. (2016) paper on cybersecurity. Costa (2014) discusses the ability of big data to provide remedies in the medical field. Thus, VRBD fits into the context of existing work in big data, and this paper describes a clearer methodology to derive value from future big data projects.

15.4 Implementing the VRBD Methodology

This section describes VRBD's implementation at the manufacturing facility. This implementation uses several different artificial intelligence/machine learning techniques. The reader should recognize that VRBD is a framework, and numerous other algorithms or heuristics can be used within this VRBD methodology.

15.4.1 Implementing the Three Vs of Big Data

As mentioned, the company generates and stores substantial information. In fact, the volume of data makes traditional data analytics intractable. File sizes are so large that the act of opening the file has exhausted memory, network bandwidth, and/or processor capacity. All of this data is stored immediately, and there is not a massive dump of data at the end of a period and so velocity exists. Finally, data is reported using Windows, Linux, and Solaris operating systems with numerous different formats. Furthermore, the log files do not follow a common schema as some machines report in pounds and Fahrenheit, but others report in grams and Celsius. Thus, this manufacturing system has volume, velocity, and variety necessary for big data. Here, the focus is on the actual implementation at this manufacturing facility.

A primary problem is storing and accessing the volume of the data. The company has a distributed computer architecture with one master computer and about 100 other computers or nodes. Hadoop® is a common open source software used in big data. Hadoop® (Shvachko et al., 2010) separates large files into smaller files, which are called shards or partitions. Each shard must be small enough that it can be stored on a single node and also accessed quickly. One could view a shard as a subset of rows of a large data set. Due to accessibility issues and to reduce errors, all data is stored on multiple shards.

The velocity of the data is also handled through Hadoop coupled with MapReduce (Lämmel, 2008; Shim, 2012). As data enters the system, the master assigns different subsets of the data to different nodes. The master typically spends some time tracking each node's amount of available storage and job requests. Thus, nodes that are less utilized are assigned jobs. The master hosts the map functions and knows what data is stored on which nodes. Furthermore, the master handles all requests for data and tells the requester, which nodes have the requested data. In MapReduce, each node has access to all the data, but such requests typically pass through the master.

The variability of the data is resolved through the reduce portion of MapReduce. The idea of reduce is to perform some analysis or other steps at each node. For instance, each node has scripts to convert different data file styles or log files into a usable format. As nodes obtain shards, these script files are run and usable data is obtained. Furthermore, these scripts would also perform data consistency functions such as converting Celsius to Fahrenheit. Thus, the MapReduce model with a sufficiently large computer cluster enables a practical implementation of big data at this manufacturing center.

MapReduce also answers basic questions or optimization problems. Some common questions that the reduce function incorporates are counts, maximums, or averages. Assume one seeks the average temperature of a metal during an entire process. Each node examines all its applicable shards and sums the temperatures and determines the number of recorded temperatures for the particular part. Every node reports these two values to the master, which aggregates this information and calculates the average. If every piece of data has been stored on an equal number of shards, then this average is correct. Thus, each node does work and reports the summary (reduced version) of the data to the master. The master compiles this data and reports the desired measure.

This facility generates data with volume, velocity, and variety. The cluster configuration, Hadoop, and MapReduce make use of numerous low caliber and cheaper machines to create a powerful and reliable computing infrastructure that can store, access, and perform analysis on the big data. The next challenge is to create value from this data. The four Rs from VRBD achieve this goal and are the focus of Section 15.4.2.

15.4.2 Implementing the Four Rs

This section describes the implementation of the four Rs at the manufacturing facility. Each R is applied at the manufacturing facility and is used to help the reader understand the implementation issues. Prior to discussing each of the Rs in detail, a formal statement of the big data is provided.

Here, the data has n instances, (X,Y). The sets $X = \{x_1, ..., x_n\}$ and $Y = \{y_1, ..., y_n\}$. Each x_i has m real numbers. Each y_i has p components, and these components are not real numbers. It should be noted that any of these items can be replaced by the empty set, which signifies that the no data was reported in the particular component. To interpret this, the x_i and y_j correspond to a particular part. For instance, the ith part reaches the inspection stage; such a part may have a set of data representing the part's cost, time of production, tolerance error, weight, etc. Each of these is stored in separate x_{ij}'s. Additionally, the part may also have other items that are not expressible as a number, such as color, notes from the production line, and type of part. This categorical data is expressed in some of the y_{ik}'s.

15.4.2.1 Remove Inaccurate or Useless Data

As previously argued, not all collected data is meaningful or correct. Here it is assumed that major errors or events are easily identified by humans in a manufacturing facility. Furthermore, any immediate problems associated with these situations are already fixed. Thus, the first R, remove, eliminates data that points to obvious errors or data that is incorrect. Such data is identified as nonconforming data or data that is classified as anomalies.

The isolation forest (iForest) algorithm (Liu et al., 2008) is an effective anomaly detection algorithm that is used in this study to detect an initial set of anomalies from the data. This data indicates major errors that are assumed to have caused serious problems, which have already been resolved. iForests use a model-based method that explicitly isolates anomalies instead of profiles normal points. Thus, abnormal data can be identified faster.

Given data (X,Y), an isolation tree is built by recursively dividing the data. The focus here is on real numbers and a binary tree. First, an attribute $j \in \{1, \ldots, m\}$ is selected along with a $p_j \in \mathfrak{R}$. The data set is divided into two sets: one has all x_i's with $x_{ij} \leq p_j$ and the other has any x_i with $x_{ij} > p_j$. This creates two new data sets $(X,Y)\leq$ and $(X,Y)>$, which are also called children. Each of these data sets is partitioned in a similar fashion by selecting a new $j \in \{1, \ldots, m\}$ and a $p_j \in \mathfrak{R}$. This partitioning continues until some preset maximum depth d or the size of a partition is one.

Expanding isolation trees beyond a binary tree and real numbers is straightforward. Instead of a single p_j for a category, several such real numbers can be selected. Thus, a set is partitioned into multiple sets of "children." Similarly, one could expand from numbers to categorical data in Y. The basic process is the same, but now, the split may involve such things as rework or original, produced on machine, and key words in a report.

Anomalies are identified by the leaf nodes. If an isolation tree has a single element in one of its leaf nodes, than that data point represents an anomaly. Similarly, a partition that has a limited number of data points in a node may also represent several anomalies. That is, a binary isolation tree should have at least a certain number of elements in each child or this partition may indicate abnormal points. There are some standard representation for this estimated partition size given randomly generated data. The idea sets $c(n) = 2H(n-1) - (2(n-1)/n)$, where $H(i)$ is the harmonic number, which can be estimated by $\ln(d)+0.5772157$ (Euler's constant). As $c(n)$ is the average of $h(x)$ given n, and the anomaly score s of an instance x is defined as $s(X, n) = 2 - (E(h(X))/c(n))$. At depth d, if there exists at most $s(X, n)$ points in a particular node, then these data points indicate abnormal data and should be removed. Thus, the isolation tree is finding isolated or abnormal data.

Expanding an isolation tree into an iForest requires training and learning. The idea is to create multiple isolation trees and average them to create an iForest. After this learning period, this iForest is the model to analyze incoming data.

In practice, one may use the aspects of the data or the system to determine the attributes that create the partitions. The associated partitioning criteria is also frequently preassigned. Even in these situations, it is still important to have at least a few random branches as these may identify some previously undiscovered nonconforming data points.

The speed of iForests is one of its greatest attractions. Each partition requires minimal effort, and applying any single data point to the iForest results in $O(d)$ effort where d is the maximum depth of the iForest. Furthermore, the partition happens with minimal effort as it only requires a comparison and moving the pointer to the data of the appropriate set. Thus, the storage requirements for iForests are also extremely small.

Before delving into an example of anomaly detection with iForests, it is import to compare this technique to another technique. Traditional statistical process control (SPC) (Grasso et al., 2015) chart is a typical manufacturing tool used to identify outliers. SPC charts need the data to be normally distributed or else inferences may not be correct. For non-normal data, individual data samples must be approximated. SPC charts are good only for single "response" usually plotted along one "predictor," i.e., time. Additional predictors can be incorporated by normalization, but this is very limited and reduces the data quality.

The proposed system was prototyped on a manufacturing lot scheduling system. The same data is depicted in Figures 15.1a and b. Each circle represents a particular part for a staging system, and there are over 20,000 such points.

Figure 15.1a shows the data analyzed through a standard control chart plot with three standard deviations away from the mean. Data inside the box of three standard deviations is typical, and anything outside is out of control and labeled as an anomaly. Observe that the upper right corner of this box is almost empty. Clearly, these data points should have been highlighted as abnormal, but the control chart method did not identify these data points as outliers. Additionally, numerous points near (250k, 7k) would be labeled out of control, when they are truly similar to many other points.

Figure 15.1b uses iForests to identify outliers. The normal points are indicated by solid gray circles, and the anomalies are highlighted in as black hollow circles. Obviously, the iForest identifies the nature of underlying data without user input. One may question the black circles near the borders. These happen to be some of the most extreme data points in a particular region. Thus, iForests appear to have identified the anomalies better than standard control charts.

Since the goal is to replicate a human's interpretation, one can easily see that iForests did a better job than SPC charts at recognizing anomalies in the data. Furthermore, iForests are extremely fast and allow the recognition of outliers to happen in real time. These outliers are discarded from the data set, and the focus now shifts toward recognizing problems.

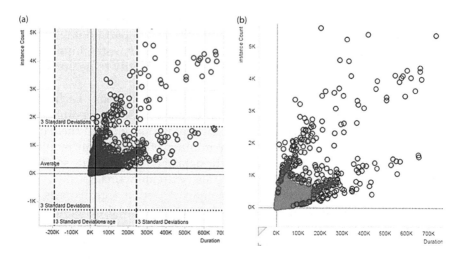

FIGURE 15.1
(a) Control chart with three σ for the manufacturing facility. (b) iForest results for the manufacturing facility.

15.4.2.2 Recognize the Problem

Recognize step's primary goal is to either identify a problem or validate that none exist. Since the abnormal data has been discarded, any problem recognized by this step is typically not identified by employees and the problem still exists. In this chapter, principal component analysis (PCA) performs the recognize step. PCA (Wold et al., 1987) has been widely used in many sectors (Vargas et al., 2018; Hu et al., 2017; Kant and Sangwan, 2014).

PCA is a popular method for dimensionality reduction. PCA requires a set of real numbers, and so this analysis is only applied to the X portion of the big data. The technique consists of deriving eigenvectors or transformation vectors that project the data into new hyperplanes. The new coordinate system is calculated in such a way that the new orthogonal coordinates (principal components) lay along the direction of maximum variation. Thus, the vectors are sorted and selected such that the first vector describes the most variance and the last vector describes the least variance within the data.

For most big data, the number of observations far exceeds the number of dimensions (m). Thus, PCA terminates with m orthogonal vectors that span the m dimensional space, assuming that the convex hull of the data is full dimensional. The next step in PCA transforms the data according to these new orthogonal vectors. Now one reduces the dimension of the space to a q dimensional space. This is accomplished by deleting the last $m - q$ orthogonal vectors. Since the orthogonal vectors are selected, such that the first vector represents the most variance, this reduction has captured the bulk of the data's variance even though the data points are no longer able to be created from the reduced data set. These q vectors are called the principal components. The distance from all data points is calculated to the space defined by the principal components. If this distance is larger than the mean plus $\lambda(\sigma)$, then this data point is recognized as an outlier from PCA. The value of λ typically ranges between 1 and 3.

Figures 15.2 and 15.3 illustrate this process for the data presented in Figure 15.1. In viewing Figure 15.2, the squares are the data points eliminated by the iForest. Thus, the black circles represent the data used for PCA. Applying PCA to the black data results in two vectors, represented by the blue lines. These two vectors are orthogonal and span the space. All data points are expressed in terms of these new vectors. To diminish the number of dimensions in the data, one eliminates the second vector. Next, the distance to the first vector space is calculated, and any points that are too far away are denoted as outliers and should represent problematic data. Figure 15.3 represents this step. The black points are good points, the gray circles are points that are eliminated by iForest but would not be eliminated by the PCA vector and the black squares are outliers to both.

To help illustrate the model, the discussions of remove and recognize have been separated. However, the facility performs both steps in unison, which is similar to how individuals simultaneously treat volume and velocity.

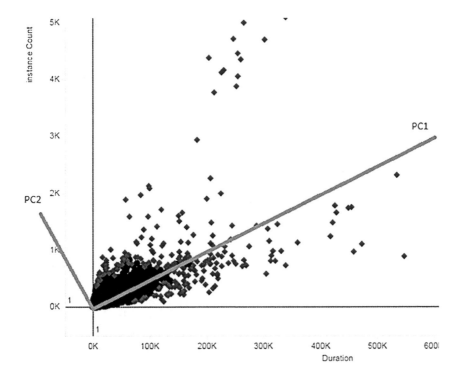

FIGURE 15.2
Graphical depiction of PCA.

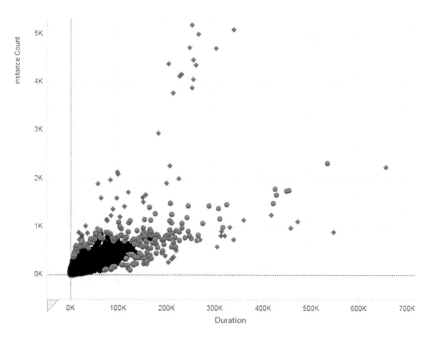

FIGURE 15.3
Results of the remove and recognize steps.

In this particular case, remove and recognize are simultaneously performed using both iForest and PCA. The iForest decreases the number of data points, and PCA then decreases the dimension for this reduced data set. All the data points are reintroduced in terms of the principal components. Any data that is classified as an outlier by both PCA and iForest is removed. Data that is classified as an outlier by one of the two methods is marked as recognized. These recognized points are gray circles in Figure 15.3 and the black points are normal. Thus, the gray circles represent potential problems and are passed to the remedy step, which is the focus of Section 15.4.2.3.

15.4.2.3 Remedy the Problem

At this point, the anomalies that have not been corrected on the shop floor have been identified. The purpose of remedy is to perform root cause analysis for these remaining outlying data. The primary idea is to cluster/categorize the data. Each cluster represents one potential problem. The data within each cluster is analyzed to identify similarities, and these similarities become suggestions for the root cause. Management receives a text or email that identifies a potential problem with one to several general suggestions related to the similarities of the data within the cluster. Management investigates the problem and attempts to fix the problem given the hints from the report.

Random decision forests (RDFs) (Ho, 1988) are implemented to identify potential remedies to the problem. RDF is an ensemble learning method for classification or clustering. The basic idea is to create a decision tree based upon properties and to partition data points according to this tree. The decisions or branching criteria are based upon some condition. Different branching criteria are compared, and the branches should be selected according to children that have the most homogenous outcomes.

For example, splitting on one criterion may result in two children with 25% and 30% of their population having a certain property. However, splitting on a different criteria results in children with 5% and 50% of their populations having the property. Clearly, the second branching criterion is better as the children are more homogenous. Once a decision tree is created, several leaf nodes should have problematic data. The clues/hints for remedying the problem are determined by backtracking through the decision tree to the root node.

A decision tree is moderately similar to an isolation tree. The primary difference is the analysis to find a better criterion to split at each node of the tree. In an isolation tree, randomness creates the branches of the trees. In a decision tree some analysis evaluates and selects the branches based upon a performance measure.

A common problem with decision trees is the over analysis of the data as the tree becomes data set specific. Random forests alleviate this problem. The idea of random forests is to sample the data set with replacement. The process samples the data and creates a single decision tree from the sample. This process is called a bag.

Creating b bags generates b decision trees. To create one random tree, these b trees are frequently averaged. In some instances, the resulting tree may select the most frequent division. The end result is a RDF, which is actually just a tree. Since each bag has some leaf nodes labeled as problems and a categorization as to why, the random decision tree also has leaf nodes labeled as problematic with potential reasons as to why. Due to the aggregation of multiple bags, the final decision tree represents all the data.

This concept of sampling from the same set of data is a classical boot strapping technique. The idea of bagging is considered training, which is a common technique in artificial intelligence. That is, the data from each bag is training the computer to develop a quality RDF. From the forest, one decision tree is created and this is the model. Thus, the computer is developing an intelligent decision tree based upon the data with no user help.

Applying RDFs to the facility is straightforward. Each morning, data from the previous day is sampled to create a random forest. Within a few minutes, the day's random decision tree is created. Data from the facility floor is processed through the decision tree. If a data points ends at a problematic leaf node, then a report could be generated. However, such a process created far too many reports, and managers never had sufficient time to investigate. Instead, if two or more data points stop at the same "problematic" leaf node over a short period of time, then the manager receives a report. This last correction enables managers to receive a few reports a day. Consequently, managers are able to investigate and solve a few problems each day.

To illustrate this remedy process, the facility desired to decrease the time parts spent in the factory. So random decision trees were evaluated based upon cycle time. Some of the problematic leaf nodes from these RDF are shown in Figure 15.4. Observe that these charts all have a reason/comment, which provides the manager with ideas on how to fix the issue. In Figure 15.4a, an alternate tool is available and the manager identified that better scheduling between machines would help resolve the issue. Changes made in evaluating a lot size resulted in a prioritization strategy for certain lots (Figure 15.4b). Multiple enhancements made in lot splitting logic for better routing are shown in Figure 15.4c and d.

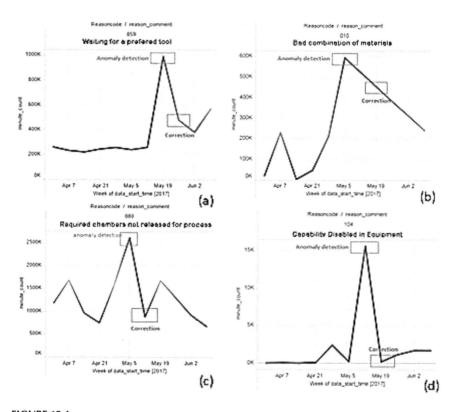

FIGURE 15.4
Results of using an RDF algorithm to remedy problems. (a) Waiting for a preferred tool; (b) bad combination of materials; (c) required chambers not released for process; (d) capability disabled in equipment.

15.4.2.4 Repeat

As previously stated, big data's three Vs are frequently augmented by a fourth V, value. In a similar fashion, the three Rs have a fourth R, repeat. The first three Rs obtain value from big data, and this fourth R relates to the frequency that big data can lead to meaningful change. Although repeat is not necessary to obtain value from big data, it does enable a large manufacturing facility to find and fix numerous problems through big data.

The most important aspect of repeat is the algorithms implemented to handle the big data projects. If the algorithms are theoretically or computationally slow, then finding and identifying problems cannot be done in a timely manner and the manufacturing facility does not rapidly react to problems.

Observe that each of the previous four sections describes methods that are polynomial and most of the ideas are linear in nature. In particular, much of the computational effort to create the models is performed once a day and incoming data is processed through these models. Thus, storing, removing, recognizing, and remedying all happen quickly. In one particular instance, VRBD identified a problem within 5 min of the initial occurrence and the manager resolved the issue within 20 min. Thus, this implementation also meets the fourth R, repeat. Section 11.5 focuses on the impact of this model.

15.5 Results from the VRBD Method

The implementation of the VRBD is still in its infancy. The project is about 6 months away from full-scale adoption. Even in its infancy, this model has proved its value to the company. Prior to VRBD, people monitored metrics and messages to hopefully highlight and fix some observed issues. The primary accomplishment of VRBD occurs during an actionable event that eliminates or reduces the impact of a problem. The new system does all of this faster and also initiates remedial action. The value to the company has occurred in four classes of problems called wolves, pythons, crocodiles, and sharks. Each of these predatory profit-eating problems are discussed along with the impact of VRBD to the facility.

The facility has numerous alarms and operators quickly learn to ignore almost all of the alarms. Thus, the facility cries wolf so often that no employee listens. The bulk of these alarms are meaningless and are labeled as a nonwolf alarm. However, VRBD recognized a nonwolf alarm for running out of a reusable material as a problem. When this alarm occurs, the reusable material that is picked up by an automated vehicle is running low. Since no employee lifts an extremely light container and the machine's alarm is being ignored, the system frequently ran out of the reusable material. This alarm is now labeled as a wolf alarm. Employees are trained on the importance of this one particular alarm and appropriate actions to take when the alarm goes off. VRBD in conjunction with management identified and remedied this problem, eliminating the wolf's consumption of productive time.

Recently, pythons were introduced in to the Florida Everglades (Dorcas et al., 2012) and have become an invasive species that is changing the natural environment. The ecosystem of modern manufacturing is in a constant state of change, and invasive species may create serious problems. Dozens of separate systems interact with each other and hundreds of engineers constantly make changes to these systems. These newly introduced changes can behave as pythons and constrict the facility's efficiency. These new activities cause both anticipated and unanticipated disruptions. The company's policy requires either a sent email or memo to notify employees of the coming change. Most employees glance over these warnings or do not fully understand the impact. In many situations, arguments occur over the machine's availability, priority, etc. VRBD identified several situations where new changes caused numerous problems. Upon investigation, the arguments were counterproductive as the desired machine frequently sat idle until the resolution of the disagreement. New policies require that the changes be noted on a shared online calendar with details regarding the time of change and the systems that may see an impact. This helped system owners investigate issues faster by narrowing down the time frame when the issue happened and what was changed during that period of time. Thus, problems from newly invasive species, such as pythons, are not constricting the productivity of the center.

Crocodiles stay concealed under the surface of the water until a sudden voracious attack captures its prey. Similarly, a manufacturing center can have issues festering just below the radar. These issues eventually spring to life causing a major event or crocodile problem. VRBD can identify this class of problems. For instance, there was an equipment alarm stating that the machines' motors are running at speeds faster than designed. The operators are in a different location than the control room, and no one recognized the problem. Furthermore, this particular alarm is a nonwolf alarm, and no employee is trained to look at it or what to do if it is alarming. VRBD recognized this as a problem and sent a report to the supervisor. Immediate action was taken and the machine did not break, nor did an accident occur. Had either event occurred, the crocodile's death roll would have consumed repair cost and lost productivity.

Great white sharks migrate across oceans through borders and near various countries (Bonfil et al., 2005), striking fear and potentially harming tourism in numerous areas. The complexity of modern manufacturing frequently requires the interaction between numerous systems. Shark problems swim across multiple departments and systems wreaking havoc due to the complexities of interaction between the systems. One shark problem identified and fixed by VRBD involved automated guided vehicles. These vehicles move material and seek to devise the best material management and placement strategy. The travel of the vehicles is passed to an intermediate enterprise management system that coordinates transactions between multiple systems. The information to move material is sent to the vehicle management software, which tries to find the best vehicle route. Several systems are interacting, and without drivers, no one knows if the vehicles are efficient. VRBD identified an abnormal level of movements for a particular material. On further investigation, the issue started with a typo in the naming of a zone. The incorrect name was sent to an intermediary system with no errors. This name went to the vehicle routing system without a problem. The vehicle routing system could not find the location and kept moving parts because these parts were never stored appropriately. Fixing the problem was trivial, and the solution shut the shark's jaws that were creating a feeding frenzy of extra work for the vehicles.

15.6 Conclusions and Future Work

This chapter introduces the VRBD method. VRBD moves big data from the virtual world into the reality of increased productivity. The primary three Vs of big data are volume, velocity, and variety. To create value, the common fourth V, the authors present the three Rs: remove, recognize, and remedy. The fourth R is repeat and focuses on the ability to apply the three Vs and three Rs rapidly.

An electronic manufacturing facility implemented VRBD by using several heuristics. MapReduce solved the problems from the three Vs. iForests, PCAs, and RDFs create fast and efficient methods to remove, recognize, remedy, and repeat. Some examples demonstrate the effectiveness of these techniques, and how it to solved problems classified as wolves, crocodiles, pythons, and sharks. These predatory problems are no longer diminishing the efficiency of the facility thanks to VRBD. The authors recommend that big data projects implement the VRBD model to improve the company's productivity.

VRBD creates numerous future research topics. Since this project is in its infancy, determining the full extent of VRBD's positive impact on this manufacturing facility is a highly anticipated future research topic. Another important question revolves around other algorithms, heuristics, or techniques to apply the remove, recognize, and remedy steps. Research should also focus on a broad implementation of VRBD. In other words, is VRBD appropriate for big data projects in nonmanufacturing sectors?

References

Armbrust, M., Fox, A., Griffith, R., Joseph, A. D., Katz, R., Konwinski, A., & Zaharia, M. (2010). A view of cloud computing. *Communications of the ACM, 53*(4), 50–58.

Assunção, M. D., Calheiros, R. N., Bianchi, S., Netto, M. A., & Buyya, R. (2015). Big data computing and clouds: Trends and future directions. *Journal of Parallel and Distributed Computing, 79,* 3–15.

Bonfil, R., Meÿer, M., Scholl, M. C., Johnson, R., O'brien, S., Oosthuizen, H., & Paterson, M. (2005). Transoceanic migration, spatial dynamics, and population linkages of white sharks. *Science, 310*(5745), 100–103.

Bowersox, D. J., Closs, D. J., & Cooper, M. B. (2002). *Supply Chain Logistics Management* (Vol. 2). New York: McGraw-Hill.

Chinta, M. (2016). *U.S. Patent No. 9,367,703.* Washington, DC: U.S. Patent and Trademark Office.

Cohen, E., & Shenker, S. (2002, August). Replication strategies in unstructured peer-to-peer networks. In *ACM SIGCOMM Computer Communication Review* (Vol. 32, No. 4, pp. 177–190). ACM.

Costa, F. F. (2014). Big data in biomedicine. *Drug Discovery Today, 19*(4), 433–440.

Dorcas, M. E., Willson, J. D., Reed, R. N., Snow, R. W., Rochford, M. R., Miller, M. A., & Hart, K. M. (2012). Severe mammal declines coincide with proliferation of invasive Burmese pythons in Everglades National Park. *Proceedings of the National Academy of Sciences of the United States of America, 109*(7), 2418–2422.

Easton, T., & Singireddy, A. (2008). A large neighborhood search heuristic for the longest common subsequence problem. *Journal of Heuristics, 14*(3), 271–283.

Fan, J., Han, F., & Liu, H. (2014). Challenges of big data analysis. *National Science Review, 1*(2), 293–314.

Garcia, S., Guarino, D., Jaillet, F., Jennings, T., Pröpper, R., Rautenberg, P. L., & Davison, A. P. (2014). Neo: an object model for handling electrophysiology data in multiple formats. *Frontiers in Neuroinformatics, 8,* 10.

Garey, M. R., & Johnson, D. S. (2002). *Computers and Intractability* (Vol. 29). New York: W. H. Freeman.

Grasso, M., Colosimo, B. M., Semeraro, Q., & Pacella, M. (2015). A comparison study of distribution-free multivariate SPC methods for multimode data. *Quality and Reliability Engineering International, 31*(1), 75–96.

Ho, T. K. (1998). The random subspace method for constructing decision forests. *IEEE Transactions on Pattern Analysis and Machine Intelligence, 20*(8), 832–844.

Hu, K., Zhang, Z., Wang, F., Fan, Y., Li, J., Liu, L., & Wang, J. (2017). Optimization of the hydrolysis condition of pretreated corn stover using trichoderma viride broth based on orthogonal design and principal component analysis. *BioResources, 13*(1), 383–398.

Ji, C., Li, Y., Qiu, W., Awada, U., & Li, K. (2012, December). Big data processing in cloud computing environments. In *Pervasive Systems, Algorithms and Networks (ISPAN), 2012 12th International Symposium on* (pp. 17–23), Taipei, Taiwan. IEEE.

Kant, G., & Sangwan, K. S. (2014). Prediction and optimization of machining parameters for minimizing power consumption and surface roughness in machining. *Journal of Cleaner Production, 83,* 151–164.

Lämmel, R. (2008). Google's MapReduce programming model—Revisited. *Science of Computer Programming, 70*(1), 1–30.

Laney, D. (2001). 3D data management: Controlling data volume, velocity and variety. *META Group Research Note, 6,* 70.

LaValle, S., Lesser, E., Shockley, R., Hopkins, M. S., & Kruschwitz, N. (2011). Big data, analytics and the path from insights to value. *MIT Sloan Management Review, 52*(2), 21–32.

Lee, M., Almirall, E., & Wareham, J. (2015). Open data and civic apps: First-generation failures, second-generation improvements. *Communications of the ACM, 59*(1), 82–89.

Liu, F. T., Ting, K. M., & Zhou, Z. H. (2008, December). Isolation forest. In *Data Mining, 2008. ICDM'08. Eighth IEEE International Conference on* (pp. 413–422). IEEE.

Marr, B. (June 20, 2016). What everyone must know about industry 4.0. Forbes, Available at www.forbes.com/sites/bernardmarr/2016/06/20/what-everyone-must-know-about-industry-4-0/#76e1eed8795f. Accessed June 7, 2018.

McAfee, A., Brynjolfsson, E., & Davenport, T. H. (2012). Big data: The management revolution. *Harvard Business Review, 90*(10), 60–68.

Milliken, G. A., & Johnson, D. E. (2001). *Analysis of Messy Data, Volume III: Analysis of Covariance.* Chapman and Hall/CRC.

Raghupathi, W., & Raghupathi, V. (2014). Big data analytics in healthcare: Promise and potential. *Health Information Science and Systems, 2*(1), 3.

Sagiroglu, S., & Sinanc, D. (2013, May). Big data: A review. In *Collaboration Technologies and Systems (CTS), 2013 International Conference on* (pp. 42–47). IEEE.

Shim, K. (2012). MapReduce algorithms for big data analysis. *Proceedings of the VLDB Endowment, 5*(12), 2016–2017.

Shvachko, K., Kuang, H., Radia, S., & Chansler, R. (2010, May). The Hadoop distributed file system. In *Mass storage systems and technologies (MSST), 2010 IEEE 26th symposium on* (pp. 1–10). IEEE.

Smit J., Kreutzer, S., Moeller, C., & Calrberg, M. (2016). Industry 4.0 Policy Department A: Economic and Scientific Policy, European Parliament, 1–94, Available at www.europarl.europa.eu/RegData/etudes/STUD/2016/570007/IPOL_STU(2016)570007_EN.pdf. Accessed June 7, 2018.

Vargas, J. M., Nielsen, S., Cárdenas, V., Gonzalez, A., Aymat, E. Y., Almodovar, E., & Romañach, R. J. (2018). Process analytical technology in continuous manufacturing of a commercial pharmaceutical product. *International Journal of Pharmaceutics, 538*(1–2), 167–178.

Veeramachaneni, K., Arnaldo, I., Korrapati, V., Bassias, C., & Li, K. (2016, April). AI^ 2: Training a big data machine to defend. In *Big Data Security on Cloud (BigDataSecurity), IEEE International Conference on High Performance and Smart Computing (HPSC), and IEEE International Conference on Intelligent Data and Security (IDS), 2016 IEEE 2nd International Conference on* (pp. 49–54). IEEE.

Wold, S., Esbensen, K., & Geladi, P. (1987). Principal component analysis. *Chemometrics and Intelligent Laboratory Systems, 2*(1–3), 37–52.

Wu, L., Barker, R. J., Kim, M. A., & Ross, K. A. (2013, June). Navigating big data with high-throughput, energy-efficient data partitioning. In *ACM SIGARCH Computer Architecture News* (Vol. 41, No. 3, pp. 249–260). AC.

Zaharia, M. (2016). *An Architecture for Fast and General Data Processing on Large Clusters.* San Rafael, CA: Morgan & Claypool.

16

Cyber-Physical Real-Time Monitoring and Control: A Case Study of Bioenergy Production

Amin Mirkouei

University of Idaho

CONTENTS

16.1 Cyber-Physical Systems

Our world is constantly changing and becoming progressively complex. The fourth industrial revolution, termed Industry 4.0, integrates mechanical inventions with cyber-enabled tools to control the processes and becomes increasingly capable of advancing the production processes and manufacturing systems (Monostori et al. 2016; Hansen and Mirkouei 2018). The motivation behind the Industry 4.0 lies in addressing energy consumption in the various sectors (e.g., industrial, commercial, and transportation), as well as inherent system failures and constraints of the existing methods in machine monitoring and data-driven process planning (Lee 2008). The most critical gaps in evaluating energy cyber-physical systems (CPSs) are due to the limited access to raw data, inadequate data extraction modules, and lack of standardized post-processing techniques to identify why an operation failed or productivity was lost (U.S. DOE 2013; Ford 2014; Roco 2016). Therefore, new advancements in automation, sensing, communication, and data collection can bridge the gaps and provide ground-breaking opportunities for future research and growth (Popovic, Kezunovic, and Krstajic 2015; Tariq et al. 2014; Clarens and Peters 2016; Han, Huang, and Ansari 2013; Zodrow et al. 2017). This section highlights the challenges of introducing CPS in manufacturing, as well as creates opportunities to provide a tangible source of data that other researchers may use to develop and validate smart-manufacturing technologies.

16.1.1 Data Analytics and Energy Prediction

Advanced data analytics collect data from various entities, analyze them based on the history of data, and make a decentralized decision after understanding the patterns and developing the prediction models. Based on the disparate nature of technologies and inherent complexity associated with cyber-physical infrastructure, it is not surprising that few works have been done to integrate CPS with existing manufacturing processes. Mainly, the intersection of energy production and CPS requires additional investigation to advance the developed computational methods, using machine learning (ML) and artificial intelligence (AI) techniques (Jha et al. 2017; Black et al. 2016). For example, support vector machine (SVM) for massive datasets with various data types can be applied as data-driven techniques to work side by side with humans (Shang, Huang, and You 2017; Connelly et al. 2016). Thus, conducting such evaluations requires an integrated performance measurement method, analytical models, metrics, and computational tools for decision support.

In evaluating production processes, many requirements enter into decision-making, such as different resources, throughput, and process conditions. Data analytics consist of two primary modules: (i) data extraction agent and (ii) knowledge extraction agent, presented in Figure 16.1. The data extraction agent retrieves raw data, such as process parameters, from target operations and performs post-processing to develop meaningful training databases without requiring any insights of the outputs. The collected data needs to be correctly contextualized and classified to identify those parameters that possibly affect the results, such as yields and product quality (Bhinge et al. 2017).

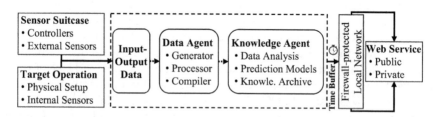

FIGURE 16.1
Data analytic components.

The raw data can be categorized into three groups—direct, indirect, and simulated data (Figure 16.2)—with the assistance of R, an open-source language and environment for statistical computing and graphics (Ihaka and Gentleman 1996). The knowledge extraction agent investigates the processed data and finds the patterns between inputs and outputs after each experiment, using the artificial neural network (ANN) and SVM. The learning algorithms analyze training datasets to predict the outputs from given inputs and develop hypothesis functions to predict future values (Mirkouei and Haapala 2014). For the production processes, several parameters can be continuously updated and affect real-time prediction models with new measurement data. The developed models can find optimal process parameters to improve energy systems.

A robust, accurate energy prediction model can be developed, utilizing generated training datasets from a set of numerous scenarios, data-driven algorithms, such as the SVM with the assistance of Python, an open-source, high-level programming language (Van Rossum 2007). The use of data-driven energy prediction algorithms is more accessible and feasible to implement than traditional practices. Previously, only empirical estimates from physics-based modeling, using physical laws, were available to account for predicting energy consumption (Bhinge et al. 2017). The response outputs achieved from real-time optical monitoring and in-line analysis can be used to develop correlations between energy consumption and various parameters of each operation (Mirkouei, Silwal, and Ramiscal 2017).

The concept of process monitoring and controlling can be extended to any parameters that affect energy consumption. Energy prediction modeling includes three major steps (Figure 16.3):

1. Understand energy consumption patterns across the energy systems.
2. Determine different operational strategies and the effects on energy consumption.
3. Derive the most effective process and energy-efficient strategy.

Rational improvements will be accomplished by effective process configurations and the trade-off between quality and yield, using computational prediction models along with real scenarios. Next section discusses coupling

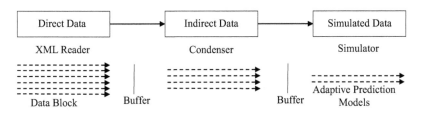

FIGURE 16.2
Data processing architecture.

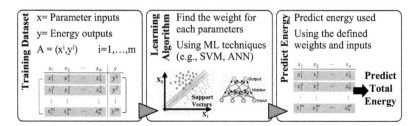

FIGURE 16.3
Energy prediction architecture.

mechanical inventions with cyber-enabled tools to (i) understand and visualize interactions between multi-scale processes and (ii) control the operations and become increasingly capable of advancing the production processes to an extent not done previously.

16.1.2 Cyber-Physical Real Monitoring and Control

In today's manufacturing economy, decision makers are looking for real-time data and information to evaluate the key parameters, gain insights, and make the best decision, while ensuring their resources are properly utilized and operated. Monitoring the reaction mechanisms and detailed kinetics, quality and yield relations, and prediction of energy consumption enables the operation process toward desired productivity and yield rates, using generated data. Real-time monitoring and analysis provide novel insights about the correlation of the necessary parameters and measurements (Hutchinson and Lee 2017).

During the past three decades, engineers have developed tools and methods for controlling the physical processes, and simultaneously, computer scientists have developed realistic methods for cyber systems (Liu et al. 2017). However, there is a considerable gap between the cyber and physical worlds, where information and real materials are transformed and exchanged. The CPS, as the fourth major industrial revolution after first three revolutions (i.e., mechanization, electrification, and automation), embraces physical components and advanced cyber-based technologies (e.g., high computation power, remote operations, interoperability standards, and communication protocols) to locally control systems and analytics (Monostori et al. 2016), as well as optimize operation performance metrics, such as accuracy, security, and reliability (Shang, Huang, and You 2017; Lee 2015). The CPS have significantly changed and exhibited more flexibility by utilizing the recent innovations. For instance, advanced CPS utilize low-cost, small-size transistors that offer more flexibility and availability of electromechanical sensors to improve performance measurement in semiconductor manufacturing industry (Gao et al. 2015; Teti et al. 2010; Lee, Bagheri, and Kao 2015). Figure 16.4 depicts simple CPS architecture, each part striving to maximize its own inherent objectives (Monostori et al. 2016).

FIGURE 16.4
Cyber-Physical System (CPS) architecture.

The training datasets developed earlier will use standard training algorithms, such as back propagation to expand predictive models. The models will consist of inputs that are expected to be varied in real-world conditions, outputs that will include results from energy production and consumption, and product throughput. Real-time process evaluations with advanced statistical, data-driven techniques capture valid process-level parameters, as well as investigate interrelationships at multi-commodity levels.

Sensing, real-time extraction and visualization of data, fast transform analysis, and cloud-based backend computing facilitate production developments and assessments in many niche applications, as well as promote process efficiency, clean energy solutions, and cross-cutting benefits (Hussain et al. 2014; Sztipanovits et al. 2014; Lu et al. 2015; Lee et al. 2012). Decision makers in different levels (i.e., operational, tactical, and strategic) will benefit from integrated resources and operations, as well as adaptive methods. The implications of CPS will be in various realms affecting STEM (science, technology, engineering, and mathematics) disciplines, such as intelligent cyber-infrastructure, advanced networking technology, and smart manufacturing (Pal and Vaidyanathan 2015; Manic et al. 2016; Mirkouei, Bhinge, et al. 2016; Wang et al. 2017; Zodrow et al. 2017).

CPS address lack of data and practical methods, along with associated deficiencies across the systems through investigating several parameters (e.g., reaction temperatures and catalyst quantities) and variables (e.g., carbon efficiency, production yield, and product quality) under real-world conditions to extend dimensions of CPS and data-driven decision-making. Table 16.1 links research questions, hypotheses, and tasks to shed light on components and support a solution-oriented project.

The need for further investigation is increasing not only in the creation of the conceptual platform but in empirical work for specific applications that can advance monitoring of the operations and increase technical growth. Growing CPS initiatives promote sustainability and resilience of design and manufacturing through integrated cyber-physical data-enabled decision-making, which is still in nascent stages, but growing steadily with improvements in sensing technologies, interoperability standards, and data-influenced decision-making. Technology breakthroughs are essential in addressing production challenges, such as process efficiency and productivity.

TABLE 16.1

Overview of Research Questions Linked to Hypotheses and Research Tasks

Research Question	Hypothesis	Task
What are the baseline, status quo interactions between operation parameters and production process entities? What are the potential approaches and tools?	Existing intricacies are not well understood; elucidation of the effects of parameters is critical to maximizing effective management of pervasive data and extracted information to meeting market needs.	Real-time process monitoring, including optical monitoring, inline analysis, and scenario analysis.
How can the integrated technologies and cyber-physical infrastructure be configured to become increasingly capable of advancing the production processes?	Deficiency in transforming current production to the next generation is not being met; particular attention should be placed on intelligent decision approaches due to disparate nature of technologies and inherent complexity associated with the cyber-physical infrastructure.	Data-driven decision-making, including data analytics, adaptive prediction modeling, and multi-layer cyber-physical infrastructures.
How can the interactions of technologies and stakeholders' needs be modified to boost resilience and sustainability benefits across energy systems?	Approaches developed in prior work have not been integrated to date to transform existing technologies to next generation; understanding the complex compounds and commercial viability of technologies provide a base of knowledge to scale up production and enhance sustainability benefits.	Sustainability assessment, including techno-economic analysis, life cycle assessment, and social analysis.

16.2 Biomass-Based Energy Production

According to the U.S. Energy Information Administration (EIA), an average of 7.8 million barrels of crude oil per day were imported into the United States in 2015 that represents a value of about $400 million per day (U.S. EIA 2017). Renewable sources have been suggested as part of a comprehensive strategy to cut the use of fossil oil in half by 2030, in addition to phasing out the use of coal (Union of Concerned Scientists 2012). Figure 16.5 demonstrates that biomass-based energy currently comprises the most substantial portion (47%) of renewables in the United States (U.S. DOE Energy Information Administration 2016); thus, more efficient process-based solutions can result in promoting bioenergy production. Since biomass is a critical renewable resource that can address national priorities, such as support energy security, mitigate global warming, and create domestic jobs (Sadasivam 2015), special attention should be placed on biomass-based energy production.

Energy sources from biomass, as a meaningful environmental solution, are being developed as a substitute for conventional energy sources (Agblevor et al. 2016; Heavers et al. 2013). However, bioenergy sources are unreliable

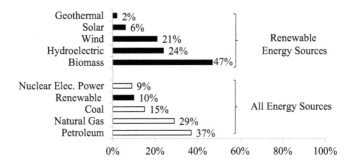

FIGURE 16.5
Total U.S. energy consumption percentage in 2016.

due to uncertain efficiency, compatibility, and profitability (Onarheim, Solantausta, and Lehto 2015). The U.S. Department of Energy (DOE) defines biomass as a renewable energy resource derived from natural materials (e.g., forest resources, energy crops, algae, agricultural residues, and animal manure) that is convertible into bioenergy (Heavers et al. 2013; Mirkouei and Kardel 2017). Biomass represents a promising renewable resource due to its abundance and low price; however, according to the U.S. EIA, over 45% of biomass feedstocks are underutilized due to immature production (conversion) technologies (U.S. EIA 2014). Earlier studies focused on evaluating biomass resources, in-depth production, and cost analyses to address upstream and downstream challenges, particularly fostering ethanol production (Cundiff, Dias, and Sherali 1997; Solantausta et al. 1992; Czernik 1994; Bridgwater and Bridge 1991). Later research focused primarily on "green" and sustainable bioenergy supply chain design and evaluated cellulosic biofuels and biodiesel (Mobini, Sowlati, and Sokhansanj 2011; You and Wang 2011; Mirkouei and Haapala 2015). Recent studies established foundational concepts in biofuels use and commercialization of new technologies by integrating various disciplines, such as algae and advanced feedstocks, strategic analysis, and cross-cutting sustainability (U.S. DOE 2016a,b; Mirkouei et al. 2016b,c; Shang, Huang, and You 2017).

Bioenergy 4.0 applies the concept of Industry 4.0 to address existing challenges (e.g., production yield, product quality, and commercialization) in the bioenergy industry and results in cost-competitive operations with low resource and processing energy required. Bioenergy 4.0 can address the limitations of current conversion process practices in bioenergy production. This section discusses the existing challenges in the traditional bioenergy production pathways with respect to the state of technology, as well as future directions for bioenergy production, coupling cyber-based datadriven decision-making with physical-based control and improvement of the operations. Finally, a cyber-physical real-time monitoring and control platform is proposed for bioenergy production from biomass feedstocks.

16.2.1 Existing Infrastructures of Bioenergy Production

Over the past 50 years, the need for more reliable, efficient, and productive production process arises in the industrial sector to address major challenges regarding high energy consumption (Helu, Hedberg, and Barnard Feeney 2017; Vogl, Weiss, and Helu 2016; Wolfe et al. 2016). Biomass-based energy conversion process challenges are associated with feedstock types, process configurations, and resources used to balance yield and product quality (Dutta et al. 2015). Carbon efficiency (the key to the commercial viability), deoxygenation (reduce oxygen content), and hydrogenation (increase hydrogen content) are three primary parameters in advancing the bioenergy conversion process (Dutta et al. 2015).

Most strategies for large-scale production of advanced biofuels focus on biochemical technology platforms, using terrestrial biomass (U.S. DOE 2016b). Lignocellulosic biomass, such as forest biomass, is an intricate matrix of polymers, e.g., cellulose, hemicellulose, starch, and lignin (U.S. DOE 2016a). Most biochemical conversion processes are, therefore, designed for a specific, limited range of feedstocks to maximize process efficiency. Biochemical technologies are not suitable for distributed production due to high capital cost and feedstock specificities as previously reported (Mirkouei et al. 2017a; Lin et al. 2016; Muth et al. 2014). In contrast, thermochemical technologies, such as pyrolysis and hydrothermal liquefaction, can be designed to be feedstock agnostic and are amenable to distributed processing, such as mobile or portable biorefinery (Mirkouei et al. 2016c; Brown, Rowe, and Wild 2013; Mirkouei et al. 2017b).

Pyrolysis is a thermochemical decomposition at 400°C–650°C temperature in the absence of oxygen, which can be grouped in two main categories: slow pyrolysis and fast pyrolysis; it differs in residence time, temperature, and heating rate (Bridgwater and Peacocke 2000; U.S. DOE 2013). Currently, fast pyrolysis is commonly used to produce bio-oil (a liquid similar to crude petroleum, with much higher oxygen and water content) and biochar (a valuable soil amendment similar to charcoal), using a wide range of algae and terrestrial feedstocks (Mirkouei et al. 2017a). The focus of this study is on fast pyrolysis because of high liquid (bio-oil) yield achieved with heating rate >1,000°C/s, resident time <2 s, and temperature 400°C–650°C. Prior studies reported that the catalytic fast pyrolysis (CFP) produces high-quality bio-oil, which needs less deoxygenation; however, the yields have been reduced (Agblevor et al. 2016; Onarheim, Solantausta, and Lehto 2015; Heavers et al. 2013; Bridgwater 2012; Black et al. 2016; Choi et al. 2016; Vasalos et al. 2016). Further details about reactors, processes, and intermediate or final products are given by Gamliel, Wilcox, and Valla (2017), Bridgwater (2012), and Badger et al. (2010).

Bio-oil varies due to complex feedstock composition and pyrolysis reactions (e.g., mixed of depolymerization, fragmentation, re-polymerization, and dehydration), which are not entirely understood. Bio-oil consists of several compounds (e.g., furan, hydroxyaldehydes, carboxylic acids, hydroxyketones, anhydrosugars, and phenolic) and has several issues (e.g., oxygen content, corrosion, viscosity, and storage characteristic) due to high oxygen-to-carbon (O/C) ratio and low hydrogen-to-carbon (H/C) ratio that indicate the quality of liquid product (Isahak et al. 2012; Evans et al. 2006). Bio-oil produced from biomass pyrolysis can potentially be used in fueling heaters, furnaces, and boilers, as well as industrial turbines, stationary dual-fuel diesel engines, and eventually upgrading to transport fuels and chemicals (Kersten and Garcia-Perez 2013; Mirkouei et al. 2017a).

CFP is touted to be one of the most promising pathways among existing, nascent thermochemical conversion technologies for cheap, local, nonfood, lignocellulosic feedstocks (Heavers et al. 2013; Pruski et al. 2016; Dickerson and Soria 2013). The presence of catalysts (e.g., Red mud, HZSM-5, $ZnCl_2$, Pt/Hbeta, and Al/MCM-41) can influence the desired modifications before the initial condensation (Agblevor et al. 2016; Pruski et al. 2016). Recent research reports transportation fuels (biofuels) and chemicals from biomass, using catalytic pyrolysis, benefit from the advantage of multifunctional catalysts, which affect various cracking and reforming reactions, and produce high-quality bio-oil by reducing viscosity, increasing H/C ratio of the final products (Agblevor et al. 2016). Current studies center on promoting catalyst effectiveness and longevity, which can address many challenges, such as liquid yield (Pruski et al. 2016). Based on the disparate nature of conversion processes and inherent complexity associated with CPS structure, it is not surprising that little work has been done to integrate CPS with existing bioenergy processes. Specifically, the intersection of CFP and CPS requires further investigation and will be covered in the section 16.2.2.

16.2.2 Future Directions of Bioenergy Production

This section centers on increasing the intelligence in biomass-based energy production process through real-time process evaluations and data-driven decision-making in order to advance bioenergy production and maximize energy consumption reduction. Additionally, a cyber-physical bioenergy production platform is proposed for CFP conversion technology, particularly bio-oil production from biomass feedstocks, such as invasive plant species. The proposed platform includes quantitative methods (e.g., ML and AI techniques) and qualitative methods (e.g., classification analysis and decision support systems), as well as a set of intelligent tools (e.g., wireless sensors and cloud-based services) for advanced data analytics to support Bioenergy 4.0 toward more efficient operations. The platform is able to examine and improve the process-level operations with CPS architecture by coupling data-driven decision-making with Internet of Things that can result in extensive sharing of information and feedback from each conversion process entity to decision makers.

Integrating cyber infrastructures and physical components of current bioenergy conversion technologies is a critical challenge to move toward Bioenergy 4.0. Bioenergy CPS decisions are influenced by dynamic business environments and suboptimal system-level solutions, which must be mitigated to accommodate industrial revolution in bioenergy production. Understanding the impact of cyber-based communication networks on physical systems aims to promote system reliability, process efficiency, and sustainability benefits (Jha et al. 2017; Baheti and Gill 2011; Graham, Baliga, and Kumar 2009; Gupta, Mukherjee, and Venkatasubramanian 2013).

Up to this time, most of the efforts have been built upon the physical-based models (Morgan et al. 2017; Woods et al. 2015), and there is an essential need in cyber-based models to construct a market-responsive conversion concept, which can produce high-value, cost-competitive products (Kezunovic et al. 2013; Gupta et al. 2011). Bioenergy production will soon be engulfed by the new advances of CPS and data-driven decision-making, which includes real-time data analytics, proactive analysis to detect problems, and predictive modeling to solve problems before happening, such as process failures or lost productivity. Thus, efficient data-driven decision-making will play a major role in the future growth of bioenergy industry (Ford 2014; Shekhar et al. 2017).

The systemic intricacies are not well understood; gaining understanding is critical not only to maximizing production efficiency and profitability but to increase the intelligence in bioenergy production through proactive analysis and predictive modeling to solve the problems before they impact the process. Advanced data-driven techniques help to capture valid process-level parameters and create data inventory, as well as improve profitability and net returns for all stakeholders by reducing the materials and processing energy used.

The cyber-physical real-time monitoring and control platform presented in Figure 16.6 is a test bed comprised of interrelated and external players who must synthesize a wealth of disparate information to make complex decisions, particularly in promoting the process efficiency and production yield. This platform relies on the latest developments in computer and data science, and manufacturing science and technology (e.g., information and

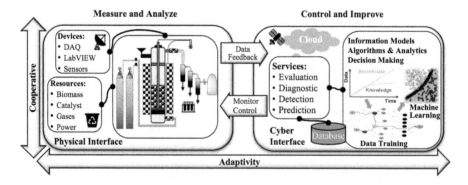

FIGURE 16.6
Cyber-physical real-time monitoring and control platform for Catalytic Fast Pyrolysis (CFP).

communication technologies), which leads to significant advantages concerning the amount of time, material, and processing energy required to generate the same amount of data by many orders of magnitude.

16.3 Conclusion

This chapter discusses the deficiency of existing bioenergy production pathways and the critical limitation of earlier studies, as well as the benefits of cyber-physical infrastructure in addressing the existing biomass-to-bioenergy supply chain challenges (e.g., market-responsive bioproducts). A scalable, multi-layer cyber-physical control and improvement platform is essential for detailed analyses (e.g., advanced data analytics and adaptive predictive analytics) and defining relationships among process parameters and variables, such as input rates of catalyst, catalyst type, and residence time. Dynamic, real-world scenarios generate a large dataset, including multiple process parameters and output variables, by creating several experiments out of each scenario. Several decision objectives, various alternatives, and pick a preferred alternative are applied, using historical and collected data to establish multiple decision criteria and to weight each criterion. Later, decision objectives are evaluated, using established criteria. These analyses provide required inputs (e.g., mass and energy calculations, size and equipment costs, cash flow, and rate of return) using various tools to evaluate productivity, capital expenditures, investment factors, and the bottlenecks of process configurations. The proposed platform herein is expected to improve the overall process performance, productivity, and flexibility through (i) diagnostic and prognostic assessment via data analytics and data-driven algorithms; (ii) resource management, quality control, and predictive maintenance; and (iii) investigating trade-offs among various factors, such as feedstock type versus yields, process conditions versus operating cost, or energy consumption versus size of pyrolysis systems. Ultimately, modernizing current bioenergy conversion pathways through adaptive cyber-physical controllers can enhance sustainability benefits across biomass-to-bioenergy supply chains at multi-spatiotemporal scales.

References

Agblevor, F.A., D.C. Elliott, D.M. Santosa, M.V. Olarte, S.D. Burton, M. Swita, S.H. Beis, K. Christian, and B. Sargent. 2016. Red mud catalytic pyrolysis of pinyon juniper and single-stage hydrotreatment of oils. *Energy & Fuels* 30, no. 10: 7947–7958.

Badger, P., S. Badger, M. Puettmann, P. Steele, and J. Cooper. 2010. Techno-economic analysis: Preliminary assessment of pyrolysis oil production costs and material energy balance associated with a transportable fast pyrolysis system. *BioResources* 6, no. 1: 34–47.

Baheti, R., and H. Gill. 2011. Cyber-physical systems. *The Impact of Control Technology* 12: 161–166.

Bhinge, R., J. Park, K.H. Law, D.A. Dornfeld, M. Helu, and S. Rachuri. 2017. Toward a generalized energy prediction model for machine tools. *Journal of Manufacturing Science and Engineering* 139, no. 4: 041013.

Black, B.A., W.E. Michener, K.J. Ramirez, M.J. Biddy, B.C. Knott, M.W. Jarvis, J. Olstad, O.D. Mante, D.C. Dayton, and G.T. Beckham. 2016. Aqueous stream characterization from biomass fast pyrolysis and catalytic fast pyrolysis. *ACS Sustainable Chemistry & Engineering* 4, no. 12: 6815–6827.

Bridgwater, A.V. 2012. Review of fast pyrolysis of biomass and product upgrading. *Biomass and Bioenergy* 38: 68–94.

Bridgwater, A.V., and S.A. Bridge. 1991. A review of biomass pyrolysis and pyrolysis technologies. In Bridgwater, A.V. and Grassi, G. (eds.) *Biomass Pyrolysis Liquids Upgrading and Utilization*, pp. 11–92. Springer, Dordrecht, Netherlands.

Bridgwater, A.V., and G.V.C. Peacocke. 2000. Fast pyrolysis processes for biomass. *Renewable and Sustainable Energy Reviews* 4, no. 1: 1–73.

Brown, D., A. Rowe, and P. Wild. 2013. A techno-economic analysis of using mobile distributed pyrolysis facilities to deliver a forest residue resource. *Bioresource Technology* 150: 367–376.

Choi, J.-S., Zacher, A. H., Wang, H., Olarte, M. V., Armstrong, B. L., Meyer III, H. M., … Schwartz, V. (2016). Molybdenum carbides, active and in situ regenerable catalysts in hydroprocessing of fast pyrolysis bio-oil. *Energy & Fuels* 30(6), 5016–5026.

Clarens, A.F., and C.A. Peters. 2016. Mitigating climate change at the carbon water nexus: A call to action for the environmental engineering community. *Environmental Engineering Science* 33, no. 10 (October 1): 719–724.

Connelly, E.B., J.H. Lambert, F. Asce, and F. Sra. 2016. Resilience Analytics in Systems Engineering with Application to Aviation Biofuels. In *2016 Annual IEEE Systems Conference (SysCon)*, Orlando, FL, 1–6.

Cundiff, J.S., N. Dias, and H.D. Sherali. 1997. A linear programming approach for designing a herbaceous biomass delivery system. *Bioresource Technology* 59, no. 1 (January): 47–55.

Czernik, S. 1994. Storage of biomass pyrolysis oils. In Proceedings of Specialist Workshop on Biomass Pyrolysis Oil Properties and Combustion, Estes Park, CO, 26–28.

Dickerson, T., and J. Soria. 2013. Catalytic fast pyrolysis: A review. *Energies* 6, no. 1: 514–538.

Dutta, A., A. Sahir, E. Tan, D. Humbird, L.J. Snowden-Swan, P. Meyer, J. Ross, D. Sexton, R. Yap, and J.L. Lukas. 2015. Process Design and Economics for the Conversion of Lignocellulosic Biomass to Hydrocarbon Fuels. Thermochemical Research Pathways with In Situ and Ex Situ Upgrading of Fast Pyrolysis Vapors. (National Renewable Energy Laboratory (NREL), Golden, CO.

Evans, R.J., S. Czernik, R. French, and J. Marda. 2006. Distributed Bio-Oil Reforming. DOE Hydrogen Program FY 2006 Annual Progress Report, Washington, DC.

Ford, S.M. 2014. *Advanced Cyber-Physical Systems for National Priorities (+$7.5 Million)*. National Institute of Standards and Technology, Gaithersburg, MD.

Gamliel, D.P., L. Wilcox, and J.A. Valla. 2017. The effects of catalyst properties on the conversion of biomass via catalytic fast hydropyrolysis. *Energy & Fuels* 31, no. 1: 679–687.

Gao, R., L. Wang, R. Teti, D. Dornfeld, S. Kumara, M. Mori, and M. Helu. 2015. Cloud-enabled prognosis for manufacturing. *CIRP Annals-Manufacturing Technology* 64, no. 2: 749–772.

Graham, S., G. Baliga, and P.R. Kumar. 2009. Abstractions, architecture, mechanisms, and a middleware for networked control. *IEEE Transactions on Automatic Control* 54, no. 7 (July): 1490–1503.

Gupta, S.K.S., T. Mukherjee, and K.K. Venkatasubramanian. 2013. *Body Area Networks: Safety, Security, and Sustainability*. Cambridge University Press, Cambridge.

Gupta, S.K., T. Mukherjee, G. Varsamopoulos, and A. Banerjee. 2011. Research directions in energy-sustainable cyber–physical systems. *Sustainable Computing: Informatics and Systems* 1, no. 1: 57–74.

Han, T., X. Huang, and N. Ansari. 2013. Energy agile packet scheduling to leverage green energy for next generation cellular networks. In *2013 IEEE International Conference on Communications (ICC)*, Budapest, Hungary, 3650–3654.

Hansen, S., and A. Mirkouei. 2018. Past infrastructures and future machine intelligence (MI) for biofuel production: A review and MI-based framework. In *ASME 2018 International Design Engineering Technical Conferences and Computers and Information in Engineering Conference*, Vol. DETC2018–86150, Quebec City, Quebec, August 26–29.

Heavers, A.D., M.J. Watson, A. Steele, and J. Simpson. 2013. Platinum group metal catalysts for the development of new processes to biorenewables. *Platinum Metals Rev* 57, no. 4: 322.

Helu, M., T. Hedberg, and A. Barnard Feeney. 2017. Reference architecture to integrate heterogeneous manufacturing systems for the digital thread. *CIRP Journal of Manufacturing Science and Technology* 19 (May 22): 191–195.

Hussain, A., T. Faber, R. Braden, T. Benzel, T. Yardley, J. Jones, D.M. Nicol, et al. 2014. Enabling collaborative research for security and resiliency of energy cyber physical systems. In *Distributed Computing in Sensor Systems (DCOSS), 2014 IEEE International Conference On*, Marine Del Rey, CA. IEEE, 358–360.

Hutchinson, C.P., and Y.J. Lee. 2017. Evaluation of primary reaction pathways in thin-film pyrolysis of glucose using 13C labeling and real-time monitoring. *ACS Sustainable Chemistry & Engineering* 5, no. 10 (October 2): 8796–8803.

Ihaka, R., and R. Gentleman. 1996. R: A language for data analysis and graphics. *Journal of Computational and Graphical Statistics* 5, no. 3: 299–314.

Isahak, W.N.R.W., M.W. Hisham, M.A. Yarmo, and T.Y. Hin. 2012. A review on bio-oil production from biomass by using pyrolysis method. *Renewable and Sustainable Energy Reviews* 16, no. 8: 5910–5923.

Jha, S.K., J. Bilalovic, A. Jha, N. Patel, and H. Zhang. 2017. Renewable energy: Present research and future scope of artificial intelligence. *Renewable and Sustainable Energy Reviews* 77: 297–317.

Kersten, S., and M. Garcia-Perez. 2013. Recent developments in fast pyrolysis of ligno-cellulosic materials. *Current Opinion in Biotechnology* 24, no. 3: 414–420.

Kezunovic, M., A.M. Annaswamy, I. Dobson, S. Grijalva, D. Kirschen, J. Mitra, and L. Xie. 2013. Energy cyber-physical systems: Research challenges and opportunities. In *NSF Workshop on Energy Cyber Physical Systems*, Arlington, VA.

Lee, E.A. 2008. Cyber physical systems: Design challenges. In *Object Oriented Real-Time Distributed Computing (ISORC), 2008 11th IEEE International Symposium*, Orlando, FL. IEEE, 363–369.

Lee, E.A. 2015. The past, present and future of cyber-physical systems: A focus on models. *Sensors* 15, no. 3: 4837–4869.

Lee, J., B. Bagheri, and H.-A. Kao. 2015. A cyber-physical systems architecture for industry 4.0-based manufacturing systems. *Manufacturing Letters* 3: 18–23.

Lee, I., O. Sokolsky, S. Chen, J. Hatcliff, E. Jee, B. Kim, A. King, et al. 2012. Challenges and research directions in medical cyber–physical systems. *Proceedings of the IEEE* 100, no. 1: 75–90.

Lin, T., L.F. Rodríguez, S. Davis, M. Khanna, Y. Shastri, T. Grift, S. Long, and K.C. Ting. 2016. Biomass feedstock preprocessing and long-distance transportation logistics. *GCB Bioenergy* 8, no. 1 (January 1): 160–170.

Liu, Y., Y. Peng, B. Wang, S. Yao, and Z. Liu. 2017. Review on cyber-physical systems. *IEEE/CAA Journal of Automatica Sinica* 4, no. 1: 27–40.

Lu, C., J. Cao, D. Corman, C. Julien, L. Sha, and F. Zambonelli. 2015. Cyber-physical systems and pervasive computing: Overlap or divergent? (PerCom Panel). In *Pervasive Computing and Communications (PerCom), 2015 IEEE International Conference*, Kyoto, Japan. IEEE, 187–188.

Manic, M., D. Wijayasekara, K. Amarasinghe, and J.J. Rodriguez-Andina. 2016. Building energy management systems: The age of intelligent and adaptive buildings. *IEEE Industrial Electronics Magazine* 10, no. 1: 25–39.

Mirkouei, A., R. Bhinge, C. McCoy, K.R. Haapala, and D.A. Dornfeld. 2016a. A pedagogical module framework to improve scaffolded active learning in manufacturing engineering education. *Procedia Manufacturing* 5: 1128–1142.

Mirkouei, A., and K. Haapala. 2014. Integration of machine learning and mathematical programming methods into the biomass feedstock supplier selection process. In *Proceedings of 24th International Conference on Flexible Automation and Intelligent Manufacturing (FAIM)*, San Antonio, TX, May, 20–23.

Mirkouei, A., and K.R. Haapala. 2015. A network model to optimize upstream and midstream biomass-to-bioenergy supply chain costs. In *ASME 2015 International Manufacturing Science and Engineering Conference (MSEC)*, MSEC2015–9355, Charlotte, NC, June 8–12.

Mirkouei, A., K.R. Haapala, J. Sessions, and G.S. Murthy. 2016b. Evolutionary optimization of bioenergy supply chain cost with uncertain forest biomass quality and availability. In *Proceedings of the IIE-ISERC*, Anaheim, CA, May 21–24.

Mirkouei, A., K.R. Haapala, J. Sessions, and G.S. Murthy. 2017a. A review and future directions in techno-economic modeling and optimization of upstream forest biomass to bio-oil supply chains. *Renewable and Sustainable Energy Reviews* 67: 15–35.

Mirkouei, A., K.R. Haapala, J. Sessions, and G.S. Murthy. 2017b. A mixed biomass-based energy supply chain for enhancing economic and environmental sustainability benefits: A multi-criteria decision making framework. *Applied Energy* 206 (November 15): 1088–1101.

Mirkouei, A., and K. Kardel. 2017. Enhance sustainability benefits through scaling-up bioenergy production from terrestrial and algae feedstocks. In *Proceedings of the 2017 ASME IDETC/CIE: 22nd Design for Manufacturing and the Life Cycle Conference*, Cleveland, OH, August 6–9.

Mirkouei, A., P. Mirzaie, K.R. Haapala, J. Sessions, and G.S. Murthy. 2016c. Reducing the cost and environmental impact of integrated fixed and mobile bio-oil refinery supply chains. *Journal of Cleaner Production* 113 (February 1): 495–507.

Mirkouei, A., B. Silwal, and L. Ramiscal. 2017. Enhancing Economic and Environmental Sustainability Benefits Across the Design and Manufacturing of Medical Devices: A Case Study of Ankle Foot Orthosis. In *Proceedings of the 2017 ASME IDETC/CIE: 22nd Design for Manufacturing and the Life Cycle Conference*, Cleveland, OH, August 6–9.

Mobini, M., T. Sowlati, and S. Sokhansanj. 2011. Forest biomass supply logistics for a power plant using the discrete-event simulation approach. *Applied Energy* 88, no. 4 (April): 1241–1250.

Monostori, L., B. Kádár, T. Bauernhansl, S. Kondoh, S. Kumara, G. Reinhart, O. Sauer, G. Schuh, W. Sihn, and K. Ueda. 2016. Cyber-physical systems in manufacturing. *CIRP Annals—Manufacturing Technology* 65, no. 2: 621–641.

Morgan, H.M., Q. Bu, J. Liang, Y. Liu, H. Mao, A. Shi, H. Lei, and R. Ruan. 2017. A review of catalytic microwave pyrolysis of lignocellulosic biomass for value-added fuel and chemicals. *Bioresource Technology* 230 (April 1): 112–121.

Muth, D.J., M.H. Langholtz, E.C.D. Tan, J.J. Jacobson, A. Schwab, M.M. Wu, A. Argo, et al. 2014. Investigation of thermochemical biorefinery sizing and environmental sustainability impacts for conventional supply system and distributed preprocessing supply system designs. *Biofuels, Bioproducts and Biorefining* 8, no. 4: 545–567.

Onarheim, K., Y. Solantausta, and J. Lehto. 2015. Process simulation development of fast pyrolysis of wood using aspen plus. *Energy & Fuels* 29, no. 1 (January 15): 205–217.

Pal, P., and P.P. Vaidyanathan. 2015. Pushing the limits of sparse support recovery using correlation information. *IEEE Transactions on Signal Processing* 63, no. 3: 711–726.

Popovic, T., M. Kezunovic, and B. Krstajic. 2015. Implementation requirements for automated fault data analytics in power systems. *International Transactions on Electrical Energy Systems* 25, no. 4 (April 1): 731–752.

Pruski, M., A.D. Sadow, I.I. Slowing, C.L. Marshall, P. Stair, J. Rodriguez, A. Harris, G.A. Somorjai, J. Biener, C. Matranga, C. Wang, J.A. Schaidle, G.T. Beckham, D.A. Ruddy, T. Deutsch, S.M. Alia, C. Narula, S. Overbury, T. Toops, R. Morris Bullock, C.H.F. Peden, Y. Wang, M.D. Allendorf, J. Nørskov, and T. Bligaard. 2016. Virtual special issue on catalysis at the US department of energy's national laboratories. *ACS Catalysis* 6(5): 3227–3235. doi:10.1021/acscatal.6b00823.

Roco, M.C. 2016. Principles and methods that facilitate convergence. In *Handbook of Science and Technology Convergence*, Bainbridge, W.S., and Roco, M.C. (Eds.) 17–41. Springer International Publishing, Switzerland.

Sadasivam, N. 2015. Economy Would Gain Two Million New Jobs in Low-Carbon Transition.

Shang, C., X. Huang, and F. You. 2017. Data-driven robust optimization based on kernel learning. *Computers & Chemical Engineering* 106: 464–479.

Shekhar, S., J. Colletti, F. Muñoz-Arriola, L. Ramaswamy, C. Krintz, L. Varshney, and D. Richardson. 2017. Intelligent Infrastructure for Smart Agriculture: An Integrated Food, Energy and Water System (May 4).

Solantausta, Y., D. Beckman, A.V. Bridgwater, J.P. Diebold, and D.C. Elliott. 1992. Assessment of liquefaction and pyrolysis systems. *Biomass and Bioenergy* 2, no. 1: 279–297.

Sztipanovits, J., T. Bapty, S. Neema, L. Howard, and E. Jackson. 2014. OpenMETA: A model-and component-based design tool chain for cyber-physical systems. In *From Programs to Systems. The Systems Perspective in Computing. Lecture Notes in Computer Science*, vol 8415, Bensalem, S., Lakhneck, Y., Legay, A. (Eds) 235–248. Springer, Berlin, Heidelberg.

Tariq, M.U., B.P. Swenson, A.P. Narasimhan, S. Grijalva, G.F. Riley, and M. Wolf. 2014. Cyber-physical co-simulation of smart grid applications using Ns-3. In *Proceedings of the 2014 Workshop on Ns-3*, 8, Atlanta, GA. ACM.

Teti, R., K. Jemielniak, G. O'Donnell, and D. Dornfeld. 2010. Advanced monitoring of machining operations. *CIRP Annals-Manufacturing Technology* 59, no. 2: 717–739.

Union of Concerned Scientists. 2012. The Promise of Biomass Clean Power and Fuel—If Handled Right.

U.S. DOE. 2013. In-Situ Catalytic Fast Pyrolysis Technology Pathway.

U.S. DOE. 2016a. 2016 Billion-Ton Report | Department of Energy.

U.S. DOE. 2016b. *Bioenergy Technologies Office—Multi-Year Program Plan*. DOE/EE-1385. U.S. Department of Energy, Bioenergy Technologies Office, Washington, DC.

U.S. DOE Energy Information Administration. 2016. Monthly Energy Review, August 2016.

U.S. EIA. 2014. *Annual Energy Outlook 2014 with Projections to 2040*, Energy Information Administration, United States Department of Energy, Washington, DC.

U.S. EIA. 2017. Total Crude Oil and Products Imports from All Countries. www.eia. gov/.

Van Rossum, G. 2007. Python programming language. In *USENIX Annual Technical Conference*, Santa Clara, CA, June 17–22.

Vasalos, I. A., Lappas, A. A., Kopalidou, E. P., & Kalogiannis, K. G. (2016). Biomass catalytic pyrolysis: Process design and economic analysis. *Wiley Interdisciplinary Reviews: Energy and Environment* 5(3): 370–383. doi:10.1002/wene.192.

Vogl, G.W., B.A. Weiss, and M. Helu. 2016. A review of diagnostic and prognostic capabilities and best practices for manufacturing. *Journal of Intelligent Manufacturing*: 1–17.

Wang, H., Y. Gao, S. Hu, S. Wang, R. Mancuso, M. Kim, P. Wu, L. Su, L. Sha, and T. Abdelzaher. 2017. On exploiting structured human interactions to enhance sensing accuracy in cyber-physical systems. *ACM Transactions on Cyber-Physical System*. 1, no. 3 (July): 16: 1–16:19.

Wolfe, M.L., K.C. Ting, N. Scott, A. Sharpley, J.W. Jones, and L. Verma. 2016. Engineering solutions for food-energy-water systems: It is more than engineering. *Journal of Environmental Studies and Sciences* 6, no. 1 (March 1): 172–182.

Woods, E.M., M. Qiao, P. Myren, R.D. Cortright, and J. Kania. 2015. Production of Chemicals and Fuels from Biomass. Google Patents.

You, F., and B. Wang. 2011. Life cycle optimization of biomass-to-liquid supply chains with distributed–centralized processing networks. *Industrial & Engineering Chemistry Research* 50, no. 17: 10102–10127.

Zodrow, K.R., Q. Li, R.M. Buono, W. Chen, G. Daigger, L. Duenas-Osorio, M. Elimelech, et al. 2017. Advanced materials, technologies, and complex systems analyses: Emerging opportunities to enhance urban water security. *Environmental Science & Technology* 51, no. 18 (July 25): 10274–10281.

17

Meta Change Management: Creating Productive and Continuous Initiatives at a Multinational Brewery Company

José Rafael Padilla Valenzuela
Heineken International

Arlethe Yarí Aguilar-Villarreal
Universidad Autónoma de Nuevo León

Daniel Ulises Moreno-Sánchez
Universidad de Monterrey

CONTENTS

17.1 Introduction: Why Change Management

According to practitioner associations such as the Association of Change Management Professionals ACMP® (2014), change management can be regarded as a process with generally agreed upon steps of an iterative and parallel nature, regardless of the methodology or approach we decide to use. Research publications, for instance, the *Journal of Organizational Change Management*, have already developed a vast body of knowledge around the discipline of change management.

The change literature has relied on contributing authors from disciplines associated with three areas: (i) psychology and sociology to know what motivates people to change, (ii) management and leadership to know how we can translate management principles to accomplish change, and (iii) engineering management and industrial engineering to develop the methods, processes, and systems by which change happens—each area providing an organizational development perspective. A full comparison of change management models is out of the scope of this work, but we provide references through the chapter for the interested reader. We recommend literature review by Al-Haddad and Kotnour (2015) as a starting point to delve into change types, enablers, methods, and outcomes.

Instead, we focus our attention on presenting four stages of transformation called the Meta Change Model, as developed and deployed at a multinational brewery company. The stages were applied for changing a business strategy for mergers and acquisitions, as well as introducing Lean Six Sigma to financial and support areas. And later institutionalized to sites worldwide as a company best practice for aiding the success rate of initiatives. Based on the acceptance that the approach has had within the company, we consider that the model could be used as a reference to create a master plan for transformation while providing guidance on how to correctly embed change in the context of companies that require the successful deployment of initiatives at sites with wide geographical dispersion.

We move on to illustrate relevant change success factors with the following anecdote from one of our colleagues. It was 6:00 am in Amsterdam, Netherlands when Jessica arrived at the gym purposeful and eager to continue, for a second year, with her new healthier life in a new city. Not a long time ago she was living in Monterrey, Mexico where her busy life was not encouraging her need to become healthier, she barely had time to go to the gym, her friends were busy as well, and they only had time to have social life after work hours. What changed then when she moved to Amsterdam, after receiving a new job offer? First, she started to meet new friends, and those friends had a very different way of behaving in terms of lifestyle than her friends back in Monterrey had. They were always eating healthy foods, riding bicycles, and going to the gym as part of their daily routine, consequently, there were hardly any unhealthy people in her new social circle. One day, Marc, one of her acquaintances, invited her to go to gym together right before work, where she could actually go walking since one of the multiple gym locations was close to her home.

What happened to Jessica that triggered this change to a new and healthier lifestyle? We may say that it was the gym proximity or perhaps the fact that she changed to a new city and it was a good opportunity to start new habits; but one factor that we know was key it was *Peer Pressure*, which is one of the many powerful elements to get people to change. When talking about organizational, social, or individual change we are talking about people behaviors and how they become habits creating a culture. If we are able to identify the factors preventing change, then we will be able to soften the same.

Change is happening everywhere all the time, either by desire or by mandate. In our industry, we are facing change in many different ways due to market conditions, new industry trends, customer's expectations, technology, processes, government, etc. Therefore, we have to find ways to come across those challenges that threaten our future as a company. One of the ways we have been very successful at managing change is via using different models to reduce resistance, accelerating transformations, and delivering more value for our organization. In this paper, we will explore these models with you and how by using the best "tricks," we can deliver superior results. Change, is often visualized in a linear pathway instead of realizing that many unexpected realities might appear (see Figure 17.1).

Success ratio for projects among companies is about 70% according to surveyed organizations (Project Management Institute, 2017), this means that on average three out of ten projects are deemed failures for not reaching their original objectives. Reasons may vary but identifying a few elements might be helpful: vision, skills, alignment, resources, action plan, and engagement. We strongly believe that as long as we fulfill all of these elements and they are well executed during a project, success will be easier to be achieved, but if any of these elements is missing during the period change takes place, failure will happen, and unpredictable outcome could be expected. See Figure 17.2 for our perspective on the success factor combinations that could happen for change management, in which only the top row assumes that success will have a higher probability of occurring when all the factors are present.

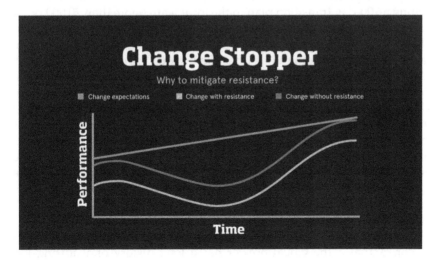

FIGURE 17.1
Change resistance chart.

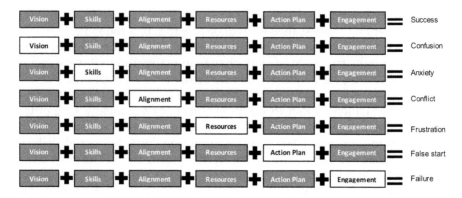

FIGURE 17.2
Success factor combinations for change management. (Adapted from Knoster, Villa, & Thousand, 2000.)

17.2 The Meta Change Model: A Path to Avoid Roadblocks and Accelerate Success

In 2010, a global brewer acquired a Mexican local brewery. At that time, there were many questions in the local operation, questions like what was going to happen with their work? Were the foreigners going to change the way the company works? Were they taking over our jobs? Those, were some of the most typical questions. Sooner than later, the global brewer started communicating a 100-day transition plan to the local team (more than 20,000 people involved), they were using the current and common communication channels the local brewery was used to. Also, they were clearly communicating that no layoffs were going to be done during that transition, they created a transition committee with the local team who was leading the change, and people were invited to participate with ideas to increase efficiency and profitability across the organization. People from the local team in Mexico were relocated to other countries as part of the people development agenda, the local management team remained with their current jobs and outsiders were only supporting with coaching, tools, and resources to help them attain their results. After only 3 months in this transition, the local teams were feeling like this was in fact a good change, it ended up being something that was not only increasing their opportunities but also helping them to grow professionally and personally.

Are these types of transitions normal across the industry? Of course, they are not. Actually, the global brewer has a long history of acquiring local brewers across the world as a way to grow inorganically. The most recent previous merger was in Europe and it was well known in the organization as a failure in terms of change management. People felt not part of the new company, they were feeling threatened by the new global teams taking over the local ones, no communication strategy was implemented whatsoever, and there were few successful projects delivering results to help the company to finance that new investment. So, what changed? Well, most of it. First, they learnt from their previous experience and made a much more comprehensive change management plan led by a local and global team.

We are calling this model the Meta Change Model since it tries deliberately to create oneness among all successful tools and models available. In Figure 17.3, the Meta Change Model is divided into four main stages and one that is common across them.

Now we will describe each model in more detail in the following pages. We begin by illustrating the creation of paradigms with a popular story involving five monkeys (Maestripieri, 2012) where a group of scientists placed five monkeys in a cage; in the middle they placed a ladder and on top of it a bunch of bananas; when one of the monkeys stepped up the ladder to grab the bananas, all others were washed with cold water. Then other monkeys tried as well and again all the other monkeys got washed with cold water, they repeated this action many times until later, when one of the monkeys wanted to grab the bananas, they were beaten up by all the monkeys, trying to prevent the cold water. After a while, no one wanted to grab the bananas. Later scientists placed a new monkey into the cage replacing one of the previous monkeys, then the new monkey wanted to grab the bananas but got beaten by the group, later, they changed the monkeys one by one until the five monkeys were all replaced. Even though the new monkeys did not know why they got beaten up by the group, when they try to grab the bananas they continue doing it because it was the way things were done in the cage.

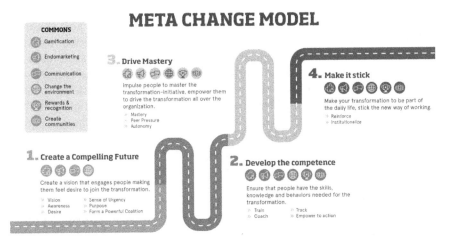

FIGURE 17.3
Meta Change Model.

17.3 The Commons: Consistent Elements on the Journey towards Change

Total Productive Management (TPM) is a company strategy used to increase productivity across all processes in the organization with the active support of all employees in a common way of working (Kerr, 2011). When our local brewer tried to implement the global TPM program in supply chain, we knew that we need to change our culture; thus, we needed to change people's behaviors. We started by developing a clear and detailed communication plan, defining the program identity (logos, iconography, colors, etc.), and then we accompanied it with a good training plan. After improvement teams were kick-started, we realized they naturally tend to compete against each other, they were trying to deliver more or better than their counterparts but there was not a support platform to do that; at that point, we stopped to think how could we leverage on this innate challenging behavior; there we came up with the idea of creating a gaming environment to make them compete in a healthy way with a strong reward and recognition program in place; we called that platform the TPM champions league, which made an analogy to the European Champions League (soccer tournament). We allowed each team to be named after their favorite football team, to use custom emblems to identify their teams, and then we created an online platform where they were able to visualize progress of each team and to compare scorecards against other teams. Needless to say, it was a success; people felt really engaged, motivated, and were delivering extraordinary results. We also established committees conformed mostly by management team members who were evaluating and, at the same time, supporting teams to achieve the outcomes expected.

In our experience, we could say that many change initiatives share common elements that allow us to soften how change is been implemented; using communication platforms such as e-mails, posters, kickoff sessions, and progress sessions. All of those are a good way to get people to know about the initiative. Gamifying the initiative is also helpful since it takes the program to a challenging level while keeping it interesting and fun. We can also reinforce the program with a rewards and recognition systems to make sure people are all on board by linking the success of the initiative to their performance indicators and then to their compensation. But even more powerful is the recognition element in here, since according to common industry knowledge, *the carrot and the stick thinking* doesn't always work.

Changing the environment is also an element of change, since you can not only hear but also see the change around an organization, and that has a far better effect on people's minds. Creating communities has a tremendous effect in conjunction with the previous elements because it creates support teams (communities) that ensure the change program is alive, they reinforce change, they recognize it, they challenge it, and they support it. Think about the communities as internal (inside the company) or external

FIGURE 17.4
Commons in the Meta Change Model.

(outside the company) parties, where you can share interest, knowledge, or just learnings. In Figure 17.4, we can find a summary of the common elements from the Meta Change Model found during the change period.

17.4 Engage: Creating a Compelling Future

Regarding culture assessment, as an anecdote, we had another of our colleagues, Juan, in a cold morning in the corporate offices; he had just opened an e-mail received in the weekend coming from HR with the recently released cultural scan results. Those included a gap analysis between the current and the desired company culture. The results are shown in the chart (Figure 17.5).

As expected, there was a gap to be closed, and they needed the whole executive team to be aware of the results and propose action plans. The executive team met and they mostly agreed that these results closely portrayed the current company culture, so they decided to start a new change program to overcome the problems. Jointly, they appointed a project leader with enough seniority and experience to help them leading the change; also, they proposed a team mostly made of representatives of every function in the organization and with the support of the change management team.

During the first team meeting, they were explained the reason they had been selected by the executive team and what was expected from them; they also felt motivated and started to configure the project plan. The team started by defining a clear vision for the team, clear success factors, resources needed, and a project plan. One of the first activities of the team, once they

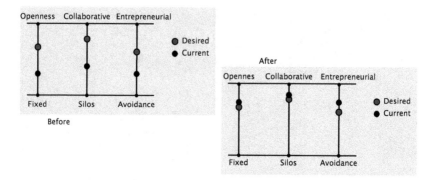

FIGURE 17.5
Aggregated cultural scan results.

had come up with a comprehensive plan to close the gap, was to do a kick-off meeting where they invited the management teams with more than 200 people in a room, they presented the *Why* they were doing that and showed the action steps, but they allowed participants to share their thoughts so that plans were flexible if changes needed to be made. They were explaining the risks of not doing this change and that only the best companies carried out such a cultural transformation. At this point, everyone was aware of the change, the only thing they did not understand was what was their role on this initiative. After a while, the project was not running according to the plan and management was not involved in the project. Only eventually they received feedback about the progress. In an emergency meeting, the project team realized they were failing and that they need to change their strategy, somehow the management team needs to be accountable for the transformation or it will not happen; but how to connect them with the management teams so they feel they own the initiative? The team soon recognized they (management teams) were not considered in the brainstorming, they did not participate with ideas on how to solve the problems, and therefore, that could be a reason for them not to be really engaged. After only weeks this single action was already showing some results, since they were part of the discussions, they were proposing actions but also helping to execute them. For example, in one of the actions, they proposed to get rid of the cubicles in the office and allow for open spaces (see Figure 17.6) with an emphasis on collaboration (see Figure 17.7). In this way, openness was not only said but also seen. This caused a huge impact in the way the company culture was perceived, and when something is perceived, it starts to feel like a reality.

Having a powerful coalition helped a lot to make sure change got implemented successfully but also connecting with the teams, helping them to find their *Purpose* to change things. One of the tools we can use to identify who are our target audiences is the stakeholder matrix (see Figure 17.8), which will help to make a stakeholder analysis and find out who we need to keep in the loop or not to avoid approaching.

FIGURE 17.6
Current open spaces at the brewery offices. (Courtesy of José Rafael Padilla Valenzuela.)

FIGURE 17.7
Current collaborative spaces at the brewery offices. (Courtesy of José Rafael Padilla Valenzuela.)

The first stage needed to drive change usually starts with having clearly identified the stakeholder's *Need*, and that need has to be conveyed in such a way that creates a sense of urgency for change; this will facilitate any transition and prevent you from trying to convince people since they are already. Not less important is to define clearly, what is expected out of this change, what is the vision, what is the picture of success; this is how success looks like in the future when we already have arrived at our destination.

In our experience, colleagues frequently ask us during training sessions about how best to create a vision. We can say that a vision represents an image of what yourself or your team aspires to become and achieve in the

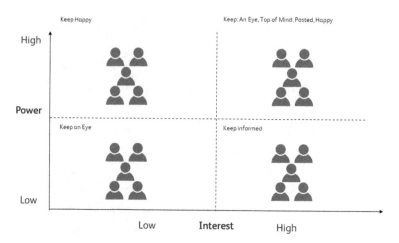

FIGURE 17.8
Stakeholder matrix.

future, the vision must be inspiring and capture people's hearts. Not just be reason driven. According to Peter Senge (2006), in order to create a powerful vision, we have to create a *Creative Tension*. The latter is explained as the gap between your vision and your current reality that creates an emotional and energetic tension that seeks to be resolved. The harder you push to create tension, the more challenging the vision is. The offspring of the vision are the objectives, which should be described in a SMART way, i.e., specific, measurable, achievable, relevant, and time bound. Once we have objectives, they can easily be translated to specific actions. In Figure 17.9, we show a summary of the first stage.

FIGURE 17.9
Transformation stage one—Create a compelling future.

17.5 Encourage: Develop the Competency

In 2014, an upgrade of the Enterprise Resources Planning (ERP) system needed to be completed, otherwise the current version would become obsolete and, in a couple of years, our ERP supplier would not be able to provide operational support. That new beverage version would help to get better insights and would run faster than the previous version with more capabilities that were very much needed in the current and ever-changing market conditions at that time. The deployment applied to more than 10,000 users across one country but in multiple locations (about 200). Also, it covered critical sales, manufacturing, and distribution processes. A team of about 140 people was organized defining first after many workshops the end state and gap to fill in each area. They started defining the training plans for key users and appointed *Super Users* from critical areas in the company; these people were selected because of the high level of performance they had shown during the years working in the company and the expertise they had in the process; but mostly because they were seen as experts within the local teams, this means that people trained by them believed in their knowledge. Special training in soft skills for the key users who would be training the trainers as well took place; they were receiving trainings in presentation skills, teamwork, and leadership, among others, besides their technical topics. These *Super Users* were given the empowerment to solve problems when they discovered them after following an established protocol. Since the deployment was planned to last about 2 years, it had to be split into waves (stages), and after each wave, a tollgate review took place to make sure the transition was going well and also to celebrate success (Kotter & Rathgeber, 2016). The training program was widespread; there were not even enough rooms available to schedule all the groups, so trailer cargo containers were prepared to supplement training facilities and were transported from one place to another according to the needs. There were for sure many challenges such as not having reliable Internet connections, computers suddenly shutting down, people not being able to attend, and training sessions needing to be rescheduled, nevertheless the change initiative was a success developing a business capability that allowed the company to lead the competition.

The training development agenda is closely aligned to the people's own career development agenda. When we focus on the latter, we have to make sure that we identify and define what competencies are required for the job and then how we can best aid the development of those.

Competencies are the observable actions performed when doing the job, which are classified into knowledge, skills, and behaviors. In order to develop these competencies, we use our learning philosophy that follows a 70/20/10 distribution. Where 70% of all learning comes from "Doing," another 20% comes from others support, and the final 10% comes from formal training.

FIGURE 17.10
Transformation stage two—Develop the competence.

In this way, we reinforce real-life experiences over other type of learning, still formal training and coaching play an important part of the development agenda.

In Figure 17.10, we show a summary for the second transformation stage.

17.6 Execute: Drive Mastery

On the first day of June in 2006, Mario was on a field trip to see how the sales force was doing the deployment of a new product recently introduced to the market. During the same day, he was accompanied by different salespersons in different routes in order for him to get a better grasp of the different market conditions; however, what he found surprised him because the level of expected standardized practices was not present. An established practice in the company was to take presale orders a day before the delivery route fulfilled those orders and collected the payment in cash or presenting a record of the transaction. Then again, he realized that the routine practices presented great differences among delivery personnel. After a full week in the market, Mario was not happy with the way the product was deployed into the market in terms of branding, promotional material, and pricing in none of the on-premise and off-premise channels; however, what concerned him the most was the lack of standardization in everything the sales and distribution forces were doing in the market, collection practices were different, also safety procedures, and so on. At that moment he knew something had to be done.

Now in the corporate offices, Mario was setting up a meeting with his team in sales to ask what could be done on that; later they decided to bring the business processes team, who reported to the IT department into the table to discuss about the options and due to the multiple implications in terms of quality, standardization, safety, compliance, and transparency, among others, a project needed to be launched. During the project planning with the leading team already in charge, they decided that in order to design new work routines (how they called them later) and then to reach more than 3,500 people in sales and more than 4,000 in distribution, they needed to bring more people to the team; then they decided to do a full-time equivalent sizing to assess the amount of people needed to deploy it; finally, more than 60 people were part of that project, which was composed of process managers, process specialists or subject matter experts, IT specialist, HR coordinators, project managers, trainers for sales and for distribution, process modelers, internal auditors, and so on.

After 1 year of team work, the daily work routines were defined, designed, and documented ready to be deployed across the organization in sales and distribution. A pilot city was selected to start with, all the team met up in one large room in a hotel with about 100 people attending the session, the presentation started, and the new process standards and daily work routines were communicated to the team using fun exercises, role playing, and games; at the end of the kickoff meeting, people were eager but also already engaged into the new change. There were many challenges during the pilot stage, people did not fully understand the training until they were coached on the field while executing the new daily routines and also at that point only the design team realized that many changes had to be done in order for the sales and distribution team to execute the routines according to design; once a newer and reloaded version was released, the full country deployment was executed. Many suggestions came into the project team hands, but they decided to continue with the deployment and update the materials on the next review phase which was planned to be done once every year. One role that was created after the pilot phase was the sales expert and distribution expert roles, which were key to ensure the quality of the implementation and was also represented a positive recognition to those frontline operators for the job they were doing that was considered superior to most of the sales and distribution force participants. They even were already role models among their peers; this helped the program to bring credibility to the new ways of working. There were still many targeted audience from the frontline that were not convinced about the change, but once they saw the experts promoting it and also their peers doing it and having good comments about it, they got convinced, mostly on the basis of *Peer Pressure*. Also, key was the decentralized approach towards the implementation plans in the countries, even though the standards were fixed, the implementation plans were not, so this allowed the countries to adapt the implementation plans and proposed good ideas to ensure a better and smoother change, this type of autonomy to the local teams helped to boost the engagement of the teams as a whole.

According to Daniel Pink (2011), human motivation has evolved from *Motivation 1.0*, in which humans are biological beings that look to fulfill their basic needs like eating, security, and sex. *Motivation 2.0* where humans respond to rewards and punishments; this works great for routine work. Again, the carrot and stick approach. And also, to *Motivation 3.0* where humans are driven to learn, create, by purpose, autonomy, mastery, and improve the world.

These three factors in *Motivation 3.0* help to generate change; Purpose: the need of human beings to feel part of something bigger, something that is meaningful for them. Autonomy: the desire to feel trusted on doing the job, not being micromanaged but empowered to propose ideas and implement them. Mastery: the need to feel we contribute to make a broader community better, becoming experts on a topic provides the motivation for change.

On the other hand, Robert Greene (2013) describes that becoming a genius is not about sudden luck but rather a relentless effort to become the best you can be by practicing over and over again until you get mastery. He even explains that the number of hours we need to become geniuses are 10,000h (approximately 5 years and a half if taking into account 8 h/day), so if you can manage to practice deliberately these many hours, you can become the best and that applies to sports, music, business, and so on.

This stage is about influencing change in the business to accelerate the transformation and deliver better and faster results. We reference the interested reader to Grenny et al. (2013) where they define a model to prosper when implementing change; the model describes six sources of influence balancing personal, social and structural motivation and ability. We have included awareness on those sources of influence in internal training sessions. In Figure 17.11, we show a summary for the third transformation stage.

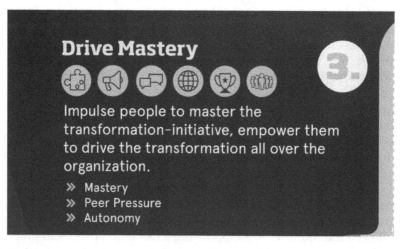

FIGURE 17.11
Transformation stage three—Drive mastery.

17.7 Conclusions

A shared objective from our experiences summed up in this chapter was the need to deliver more value for the organization. Whether that required changing the technology, cultural behaviors, or improving productivity, the teams at the brewery were able to take those changes to the next level relying in a comprehensive change management model. For us, the soft benefits of the results were evident with the updated cultural scan results that reflect an increased desire to change the game attitude, more fulfilled employees, and a collective sense of achievement. Although many companies would inevitably face many failures during their change programs, we have to embrace those opportunities as part of the learning process, failure is part of change.

The Meta Change Model expresses our openness and flexibility to adapt to new ways of working, by managing change in the smartest way possible. The model has been versatile to our needs, having the possibility to incorporate newer tools as it fits the situation.

References

Al-Haddad, S., & Kotnour, T. (2015). Integrating the organizational change literature: A model for successful change. *Journal of Organizational Change Management*, 28(2), 234–262.

Association of Change Management Professionals. (2014). Standard for Change Management.

Greene, R. (2013). *Mastery* (Reprint ed.) London: Penguin Books.

Grenny, J., Patterson, K., Maxfield, D., McMillan, R., & Switzler, A. (2013). *Influencer: The New Science of Leading Change* (2nd ed.) New York: McGraw-Hill Education.

Kerr, C. (2011, June). Zoeterwoude, 10 June 2011 TPM—The HEINEKEN Company. Retrieved February 2018, from www.theheinekencompany.com/-/media/Websites/TheHEINEKENCompany/Downloads/Download%20Center/Presentation/2011/10%20June%202011%20Heineken%20Investor%20Seminar%20Zoeterwoude%202011/Heineken%20NV%202011%20Investor%20Seminar%20Zoeterwoude%20Christopher%20Kerr.

Knoster, T., Villa, R., & Thousand, J. (2000). A framework for thinking about systems change. In R. Villa, & J. Thousand (Eds.), *Restructuring for Caring and Effective Education: Piecing the Puzzle Together*. Baltimore: Paul H. Brookes Publishing Co.

Kotter, J., & Rathgeber, H. (2016). *Our Iceberg Is Melting: Changing and Succeeding Under Any Conditions* (2nd ed.) New York: Penguin Random House.

Maestripieri, D. (2012, March). What Monkeys Can Teach Us About Human Behavior: From Facts to Fiction. Retrieved February 2018, from Psychology Today: www.psychologytoday.com/blog/games-primates-play/201203/what-monkeys-can-teach-us-about-human-behavior-facts-fiction.

Pink, D. (2011). *Drive: The Surprising Truth About What Motivates Us*. New York: Riverhead Books.

Project Management Institute. (2017). Success Rates Rise | Pulse of the Profession 2017. Retrieved February 2018, from www.pmi.org/learning/thought-leadership/ pulse/pulse-of-the-profession-2017

Senge, P. (2006). *The Fifth Discipline: The Art & Practice of The Learning Organization*. New York: Doubleday.

Section III

Postface

18

Remarks on Future Directions and Next Steps

Harriet B. Nembhard
Oregon State University

Elizabeth A. Cudney
Missouri University of Science and Technology

Katherine M. Coperich
FedEx Ground

The collaboration between industry, government, and academia is still evolving. The intent of the book is to bring together experts in their respective field to display what enables success in innovative project collaborations. Everyday examples of these collaborations are illustrated in the case studies contained in this book. There are frameworks of building teams and collaborative approaches to solve real-world issues. This publication is intended to be a launch pad for future volumes to build on.

For us, the real long-term value of the book is the repository of methods and case studies that serve as a benchmark for future collaboration. Nevertheless, it is also a learning tool for what is possible when a diverse set of individuals look for an innovative approach to solving problems. Even though we may be rooted in industrial and systems engineering (ISE), these frameworks and case studies are a testimony to what can be achieved when issues are examined through different ISE lenses.

Index

A

B